W9-CHR-124

PARTICLES AND POLICY

Masters of Modern Physics

Published Volumes

SPICE & WOLF 🕤

ISUNA HASEKURA
KEITO KOUME
CHARACTER DESIGN:
JYUU AYAKURA

TRANSLATION: JASMINE BERNHARDT

LETTERING: KATIE BLAKESLEE

OOKAMI TO KOUSHINRYOU VOL. 15
©ISUNA HASEKURA/KEITO KOUME 2017
EDITED BY ASCII MEDIA WORKS
FIRST PUBLISHED IN JAPAN IN 2017 BY
KADOKAWA CORPORATION, TOKYO.
ENGLISH TRANSLATION RIGHTS ARRANGED WITH
KADOKAWA CORPORATION, TOKYO,
THROUGH TUTTLE-MORI AGENCY, INC., TOKYO.

ENGLISH TRANSLATION © 2018 BY YEN PRESS, LLC

YEN PRESS
1290 AVENUE OF THE AMERICAS
NEW YORK, NY 10104

VISIT US AT YENPRESS.COM
FACEBOOK.COM/YENPRESS
TWITTER.COM/YENPRESS
YENPRESS.TUMBLR.COM
INSTAGRAM.COM/YENPRESS

FIRST YEN PRESS EDITION: MARCH 2018

YEN PRESS IS AN IMPRINT OF YEN PRESS, LLC.
THE YEN PRESS NAME AND LOGO ARE TRADEMARKS OF YEN PRESS, LLC.

LIBRARY OF CONGRESS CONTROL NUMBER: 2015956856

ISBNs: 978-1-9753-0011-1 (PAPERBACK)
        978-1-9753-2698-2 (EBOOK)

10 9 8 7 6 5 4 3 2 1

BVG

PRINTED IN THE UNITED STATES OF AMERICA

PARTICLES AND POLICY

WOLFGANG K.H. PANOFSKY

The American Institute of Physics

AIP Press
American Institute of Physics
500 Sunnyside Boulevard
Woodbury, NY 11797-2999

Library of Congress Cataloging-in-Publication Data

Panofsky, Wolfgang Kurt Hermann, 1919–
    Particles and policy/Wolfgang K. H. Panofsky.
        p. cm.—(Masters of Modern Physics; v. 8)
    Includes index.
    ISBN 1-56396-247-0
    1. Particles (Nuclear physics). 2. Arms control. 3. Science
and state. I. Title. II. Series.
QC793.28.P35   1993   93-14011
338.9'26–dc20   CIP

This book is volume eight of the Masters of Modern Physics series.

Contents

About the Series

Masters of Modern Physics introduces the work and thought of some of the most celebrated physicists of our day. These collected essays offer a panoramic tour of the way science works, how it affects our lives, and what it means to those who practice it. Authors report from the horizons of modern research, provide engaging sketches of friends and colleagues, and reflect on the social, economic, and political consequences of the scientific and technical enterprise.

Authors have been selected for their contributions to science and for their keen ability to communicate to the general reader—often with wit, frequently in fine literary style. All have been honored by their peers and most have been prominent in shaping debates in science, technology, and public policy. Some have achieved distinction in social and cultural spheres outside the laboratory.

Many essays are drawn from popular and scientific magazines, newspapers, and journals. Still others—written for the series or drawn from notes for other occasions—appear for the first time. Authors have provided introductions and, where appropriate, annotations. Once selected for inclusion, the essays are carefully edited and updated so that each volume emerges as a finely shaped work.

Masters of Modern Physics is edited by Robert N. Ubell and overseen by an advisory panel of distinguished physicists. Sponsored by the American Institute of Physics, a consortium of major physics societies, the series serves as an authoritative survey of the people and ideas that have shaped twentieth-century science and society.

Preface

This is a collection of papers written for nonspecialists in elementary-particle physics and in arms control, the two topics that comprise most of my technical writings. The papers are selected because of their broad interest rather than their critical substantive significance.

Let me explain how this combination of topics arose. I left Germany in 1934 at the age of 15, having thus far received a very classical education. As an undergraduate at Princeton University, my interest was deflected into science—partially as a "path of least resistance" where the language and culture shock in moving across the Atlantic Ocean had minimum impact. I then did graduate work at Cal Tech using x rays for precision measurements on natural constants. I became acquainted with elementary particle physics at the same time. Then came the war. By a series of fortuitous circumstances, I became the link between Cal Tech and Los Alamos—transferring technology developed at Cal Tech to measure shock waves from supersonic projectiles to the applications of measuring the yield of the first nuclear weapons. I participated in the Trinity Test at Alamagordo, flying in an airplane at a distance of 10,000 feet from the first nuclear explosion and having to trust the theoretical physicists who predicted that that was a safe place to be.

At Los Alamos, this work was done with Luis Alvarez—who, at that time, hatched ideas about proton linear accelerators. He persuaded me to return after the war to the Berkeley Radiation Laboratory to help him in making such accelerators a reality. These contacts at Los Alamos set me on a dual path. My principal work became accelerator design and building and experiments in elementary particle physics. I also did a great deal of teaching at all levels and wrote a graduate-level textbook. But, throughout the subsequent period, I maintained my avocation—that is, trying to clarify the frequently turgid concepts about what nuclear weapons are about and trying to work both within and without the government to get some sanity and restraints in controlling the evolution of nuclear weapons.

The work on accelerators led from the 32-MeV proton accelerator at Berkeley, to the series of electron linear accelerators at Stanford. The work on particle physics led from the initial exploration of the neutral pion at Berkeley, to many other experiments using the Radiation Laboratory's complex of accelerators. This continued at Stanford on the first experiments on inelastic electron scattering from the proton, muon pair creation, and many other particle physics experiments.

In parallel with these activities, my responsibilities increased for managing the evolving laboratories at Stanford and the construction projects that were leading to the sequence of electron linear accelerators of ever-increasing power. To some extent, these increasing managerial responsibilities led to a merging of my vocation and avocation—that is, the direct technical and scientific work on accelerators became linked to consideration of science policy as it affects decision making about science. At the same time, the concern with arms control led to my role as an advisor on how science should affect governmental policy decisions. I served on the President's Science Advisory Committee from 1959 to 1964.

As a result of the above sequence of events, "particles and policy" initially entered my work as totally separate topics, but some merging between the two followed as I became involved not only with policy applied to science, but also with science applied to policy.

The papers selected here generally emphasize a particular issue where clarification and a popular understanding of science seemed necessary. The first group of papers (1-4) covers scientific topics.

"High-Energy Electron and Photon Physics" attempts to illustrate how this type of physics complements the type of experimental work that can be done at proton accelerators. The primary, generally accepted thesis is that, since the electromagnetic interaction is well understood, results obtained through the use of electrons and photons (which do not interact directly through the strong interaction) in exploring particle structure would give results that are easier to interpret.

"Colliding Beams versus Beams on Stationary Targets" discusses the pros and cons of particle accelerators controlling both types of beams.

Again, complementarity, rather than competition, is emphasized.

"Special Relativity Theory in Engineering" was written as part of a celebration of the 100th anniversary of Einstein's birthday. Its main message is that, contrary to general public impression, special relativity is not an esoteric subject known to a few experts, but has become the basis of large fields of engineering, wherever high voltages are involved or where extremely high-precision measurements from objects orbiting in space are required.

"Particle Substructure" attempts to demonstrate the basic historical continuity between the early experiments in the 1930s observing inelastic scattering of photons on the atom and the experiments done at SLAC (Stanford Linear Accelerator Center) right after its completion on inelastic scattering of electrons on neutrons and protons. The analysis of the scattering events by x rays in the first case, and electrons in the second case, revealed the dynamics of the objects under investigation—which in early experiments was the atom and in later experiments was the nucleon. The analogy is, in fact, quite striking.

The next group of papers (5–7) constitute the transition from discussion of specific technical topics to policy questions applied to science.

"Basic Research: Curse or Blessing" was given on the occasion of the 25th anniversary and founding of DESY (Deutsches Elektronen Synchrotron) in Germany, the leading high-energy physics laboratory using electron beams in Europe. It addresses the controversy over whether the pursuit of science to its "bitter end" is a good or bad thing. The paper supports the conclusion that pursuit of basic science should remain unfettered while societal constraints should be progressively exerted as applications become nearer.

"Big and Small Science" discusses the apparent, but (in my view) not real, conflict between "big" and "small" science, and attempts to emphasize that, while the tools of many branches of science might differ, the basic motivation of the scientists pursuing the work does not.

"Technical Limits for High-Energy Proton and Electron Colliders," closely related to the previous paper, addresses the question of how big accelerators can get in any practical sense, given the limits of basic technical factors. This question relates, of course, to the previous one—namely, whether the big science–small science discussion, as it affects accelerators, will not, in fact, be settled by technical rather than policy factors.

The subsequent papers (8–12) shift to issues on the arms buildup and the efforts to control it. "Science, Technology and the Arms Buildup" addresses the concern that perceptions, rather than technical and scientific reality, frequently drive arms races, and that a firmer recognition of technical realities should lead to increased exercise of restraint in the arms buildup.

"Mutual Hostages" and "MAD versus NUTS" deal with what in my view is the unavoidable conclusion that the emergence of nuclear weapons has established a mutual hostage relationship among the world's populations. Since potential destruction is so enormous and extremely difficult to mitigate, this enormous potentially destructive power makes the world's nations hostages for peace. Both papers contain a warning that the efforts to fine-tune the use of nuclear weapons in warfare is counter-productive. The paper "MAD versus NUTS" (Mutually Assured Destruction vs. Nuclear Utilization Target Selection) attempts to illustrate the futility of failing to acknowledge the mutual hostage relationship and instead developing strategies that pretend to be able to predict, with precision, how nuclear weapons might be effective in selective, actual use in warfare.

"Misperceptions about Arms Control" was originally addressed to a large audience of accelerator specialists. Arms control is not disarmament. Rather, arms control is part of national security and is not in any way inimical to national security. In other words, arms control seeks maximum security at minimum risks and burdens of armaments.

"Arms Control, Compliance and the Law" aims to inject some clarity into the debate whether verification of particular arms control agreements is or is not sufficient. It tries to draw analogies, wherever applicable, between verification of international arms control agreements and the enforcement of domestic legislation. Its main thesis is that the nations should decide whether, on balance, national and international security would be better off with the arms control agreement in place, taking the anticipated compliance into account, than without that agreement. Demanding absolute standards of verification, that is, demanding that no violations should remain undetected and unidentified, is clearly absurd—similar to demanding that murder should remain legal since not all murderers can be brought to justice.

The last paper, "Science Advice to the President," was a contribution to a series examining a variety of attitudes on how science advice to the president should be structured in the future, following the demise under President Nixon of the President's Science Advisory Committee. The paper reflects much of my experience as a member of the President's Science Advisory Committee.

High-Energy Electron and Photon Physics: Exploring the Unknown with the Known

In the past, most high-energy physics research was carried out with proton accelerators. Such machines have been used both for the study of the interactions of the primary protons themselves and as generators of secondary unstable particles. Electron machines have been used for studies of the primary electron and photon interaction, but they have not served extensively as sources of secondary particles. This situation is changing greatly: Electron accelerators are entering the mainstream of high-energy physics, both as tools for studying the interactions of electrons and photons and as sources of secondary particles.

Studies of high-energy physics using electron and photon beams give, in many respects, much more basic information than processes initiated by protons. This is due primarily to the dominant role played in all of physical science by one of its branches—namely, quantum electrodynamics.

We have recognized several basic forces of nature, notably gravitation, the electromagnetic interaction, the nuclear strong reaction, and the nuclear weak interaction. It is fair to say that, among these at present, only the electromagnetic interaction, as described by quantum electrodynamics, can be described completely in mathematical terms: numerical agreement between experimental and theoretical results is spectacular. It is also

A talk given at National Academy of Sciences Symposium, "Revelations and Prospects in High-Energy Physics," April 24, 1968.

fair to say that all of atomic physics, all of solid-state physics, all of chemistry, and probably all of biology can ultimately be considered to be a manifestation of quantum electrodynamics; in these fields of science, electrical forces alone control the physical phenomena. Quantum electrodynamics has been demonstrated to be correct to high accuracy over the full range of distances from atomic to subatomic dimensions. A small exception to this statement is provided by a remaining unexplained discrepancy among different methods of determining the value of the fine-structure constant. It would be a result of extreme significance if quantum electrodynamics were to cease to be valid at very small distances.

I would like to state here that all experiments directed toward this question thus far have given a null result: That is, quantum electrodynamics apparently is a valid description of the electromagnetic interaction down to distances as small as 10^{-15} centimeter. I will discuss later how this question is examined experimentally and what the prospects are for pushing the frontier of our knowledge of the validity of quantum electrodynamics still further. Let me assume for the moment that quantum electrodynamics is a valid theory; it is this fact that makes high-energy physics with electron and photon probes such a unique undertaking. Because the electron, the photon, the mu meson, and the neutrino do not carry the nuclear strong interaction, and because the nuclear weak interaction is about 10^{-10} times weaker than the electromagnetic interaction, one can consider these particles as carriers of the electromagnetic interaction only when they are used as probes of the structure of other elementary particle systems. In electron and photon physics, we can therefore explore the *unknown* structures of the elementary particles and their excited states by carriers of a *known* interaction, rather than having to explore the unknown with the unknown, as is the case with proton incident beams.

The primary example of a fruitful exploration of an unknown structure using the electromagnetic interaction only is high-energy electron–proton and high-energy electron–neutron scattering. It is now known that the radius of the nucleons is on the order of 10^{-13} centimeter, and wavelengths smaller than this number are therefore required to explore nucleon structure. According to the uncertainty principle, the wavelength of exploration is inversely related to the momentum transfer to the nucleon system in a scattering process; numerically, studies of nucleon structure therefore require momentum transfers greatly in excess of 100 million electron volts divided by the velocity of light (100 MeV/c). Electron scattering on the proton has been carried out with all early electron machines; it was

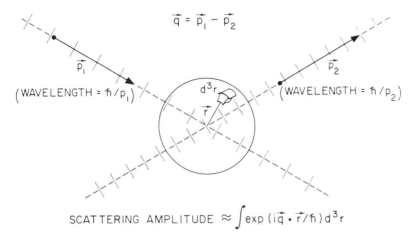

FIGURE 1. *Wave description of nonrelativistic electron–proton scattering.*

the Nobel Prize–winning work of Robert Hofstadter, using the 1 giga-electron-volt (1 GeV) accelerator at Stanford University, that first uncovered a proton structure differing from a point.

One can look at electron scattering on the proton in several alternate ways: If it is viewed as a classical diffraction process (Figure 1), the incident electron can be considered as a plane wave, each element of which scatters from a point within the nucleon that has a charge density of $\rho(r)$. The resultant scattering amplitude in a nonrelativistic language can easily be shown to be proportional to $\int \exp (iq \cdot r/h) \rho(r)d^3r$, where q is the vector difference between the incident and outgoing electron momentum. In other words, the scattering cross section of electrons on the nucleon yields the Fourier transform of the charge distribution of the nucleon. The form of the Fourier relation is a direct representation of the uncertainty principle: the larger the momentum transfer q in the exponent, the more sensitive will the integral be to variations of the charge density $\rho(r)$ for small changes in the coordinate r. In more modern terms, we look at this kind of electromagnetic process by means of a diagram in which the incident electron "virtually" emits a photon that is then absorbed by the nucleon system (Figure 2). Quantum electrodynamics then defines a formalism into which the structure of the nucleon enters in terms of "form factors," which are, in essence, Fourier transforms similar to the ones discussed earlier in this paragraph. However, the more complete theory indicates that, to describe the electromagnetic structure of the proton and neu-

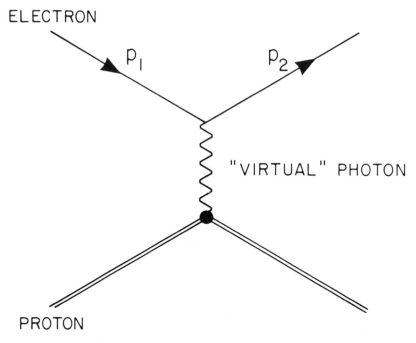

FIGURE 2. *Feynman diagram of electron–proton scattering.*

tron completely, one requires two form factors rather than one; these two numbers describe, respectively, the distribution of electric charge and the magnetic moment (that is, the current distribution) within the nucleons.

Electron-scattering experiments have now been extended at SLAC to momentum transfers as high as 5 giga-electron-volts divided by the velocity of light (5 GeV/c), which corresponds to distances on the general order of 10^{-15} centimeter. Figure 3 shows the resultant form factors obtained. Several surprising facts have emerged from these recent results: The first is that the form factors continue to decrease rapidly (as the inverse fourth power of the momentum transfer, even at these very high energies). Physically, this means that, even at this fine resolution, the proton does not exhibit a core or any other central structure that would cause the scattering amplitudes to level off at high energies, or, to be precise, at high momentum transfers.

Let me now digress briefly from discussing results to discussing experimental techniques. I would like to demonstrate that, although the style and size of carrying out high-energy physics experiments has changed

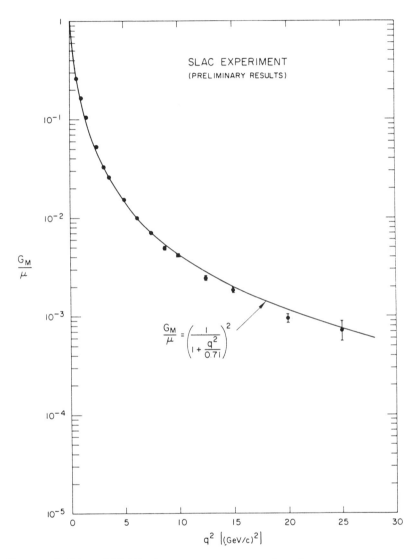

FIGURE 3. *Plot of SLAC electron–proton scattering data expressed as a "form factor."*

very much during the last decades, the basic purposes and fundamental ideas of experimentation have not. The type of experiment that I have discussed above is characterized by the fact that the yield of interesting events is small; for instance, if the instrumentation is set to examine the

FIGURE 4. *Aerial view of SLAC.*

lowest point on the curve of Figure 3, one observes only one actual recorded event per 10^{18} electrons incident on a liquid hydrogen target in
which the scattering takes place. Why is this yield so small, and how is it
measured? The answer to the first half of the question is simply that elastic scattering cross sections become exceedingly small; in fact, at the
point in question the cross section is comparable to that of neutrino reactions that, insofar as this is known, exhibit the nuclear "weak" interaction
only. The reason for this small yield is that the proton appears to remain a
large, diffuse structure, so that scattering from different parts of the structure is not likely to interfere constructively in the particular direction of
scattering. Moreover, the basic electromagnetic force is weaker than the
nuclear interaction that dominates in nuclear physics.

As for our method, we use the basic components always used in this
kind of investigation: a source of incident particles (in this case, the Stanford Linear Accelerator); a spectrometer, which sorts out the desired from
the undesired processes occurring in the target as it is bombarded by the
incident particles; and a detector followed by electronic devices designed
to analyze the data generated. Figure 4 shows an aerial view of the Stan-

FIGURE 5. *Interior view of target area with spectrometers.*

ford 2-mile machine, which is the source of the particles. The experiments in question are carried out in one of the target buildings to the south (to the lower right), toward which the beam is deflected. Figure 5 shows the spectrometer arrangements used. The incident beam strikes the liquid-hydrogen target located at the central pivot, and a magnetic spectrometer rotates around this pivot. Figure 5 shows three different instruments designed to cover different ranges of scattering angle. The spectrometers are designed and built with high precision; they use magneto-optical elements to disperse the *momentum* of the scattered particles in this vertical plane while different horizontal production *angles* are being brought into foci that are dispersed horizontally. As a result, the different particles scattered in the target are being guided along trajectories that focus inside the detector shield. A set of counter hodoscopes inside the shield then transmits this information to the data-handling system. Figure 6 shows a typical data display processed by computer: The figure shows the frequency distribution of orbits entering the shield as a function of horizontal and vertical coordinates—that is, as a function of particle scattering angle and momentum. The kinematics of the elastically scattered particles is easily recognized: The crest of this three-dimensional representation corresponds to that relation between angle and mo-

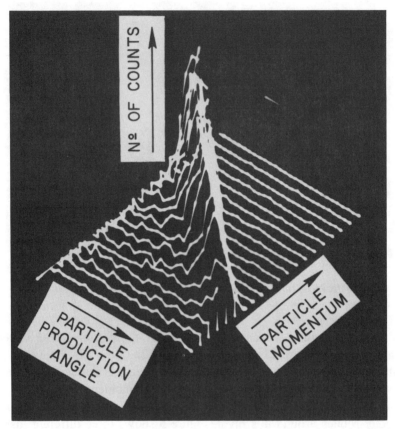

FIGURE 6. *Counts recorded in spectrometer hodoscope produced in elastic scattering. These data, which are plotted by computer, show the momentum of the scattered particle and its production angle. Note the "radiative" tail.*

mentum of the scattered electron required by the dynamics of elastic scattering. Notice also that this ridge is asymmetric toward lower momentum. The reason is that, as electrons are scattered, they may, concomitant with the scattering process, emit one or more photons (or x rays), which can reduce the momentum of the scattered particle.

In spite of the mammoth size of instruments of this kind, the precision required and realized in this work is comparable to that attained in low-energy spectroscopic studies. Energy resolution of the three instruments shown in Figure 5 is ±0.05 percent and angular resolution is about 1 minute of arc. Moreover, the flux of scattered electrons to be detected is accompanied by other particles that are not of interest in the experiments in question and that may exceed in numbers those to be measured by a factor as large as 10^6. Hence, thorough particle identification is required.

I have described only one of the many experimental techniques available in this field, and I have emphasized the similarity of such techniques with those in past practice rather than their dissimilarity. Today there are, of course, many other fundamentally new techniques in use in high-energy physics in general, and in electron and photon physics in particular, that I cannot describe here.

The experimental result discussed above relates to "elastic scattering"—that is, scattering in which the incident energy is divided between the scattered electron and the recoiling proton but in which the recoil proton remains in its *ground state*. It is now known that the proton has a "spectroscopy"—that is, it has many *excited states* above the ground state, and physicists are beginning to understand the location of the energy levels, the term assignments of the levels, the selection rules governing transition between levels, and the regularities of the level patterns. The principal lack is an understanding of the forces that establish these levels to start with and that govern the rates of transition between them. You may recall that this was precisely the situation circa 1920 with respect to atomic energy levels before the exploitation of quantum mechanics. You may also recall that one of the principal experiments shedding light on this situation was the Franck-Hertz experiment. In this experiment, electrons with the energy of a few electron volts were passed through an atomic vapor; the energy spectrum of the transmitted electrons was explored by measuring the transmitted current as a function of the applied accelerating voltage. As shown in Figure 7, the transmitted current then showed "bumps" or "breaks" wherever the energy of the transmitted beam had suffered a loss corresponding to the excitation of atomic energy levels in the vapor. This same idea can now be applied in an energy region 10^9 times higher with equally impressive results. The Franck-Hertz technique, when translated to high-energy electron physics, is called "inelastic electron scattering" and consists of examining energy spectra of scattered electrons using apparatus essentially identical to that described above for elastic scattering. Here, however, these electrons scattered on the proton have suffered energy losses corresponding to raising the proton to one of its excited states. Figure 8 shows an inelastic spectrum of this kind. You can see here clearly that this new spectroscopy

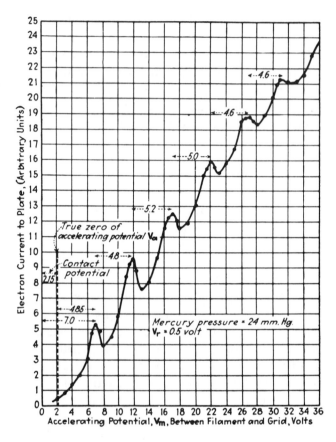

FIGURE 7. *Results of a Franck-Hertz experiment in mercury vapor. The transmission of an electron beam as a function of electron-beam voltage is shown.*

of the nucleon is real; that is, you can see the energy levels of the excited nucleon directly, as represented by the energy loss of the scattered electron. By studying the relative height of the peaks (and of the continuum following them at high excitations) as a function of momentum transfer given to the nucleon system, we can learn a great deal about the structure of the excited states. We find that the variation of inelastic scattering with increasing momentum transfer is less steep than that shown earlier in Figure 3 relating to elastic scattering. There are, in fact, theoretical speculations, now the subject of experimental study, that would predict that, if

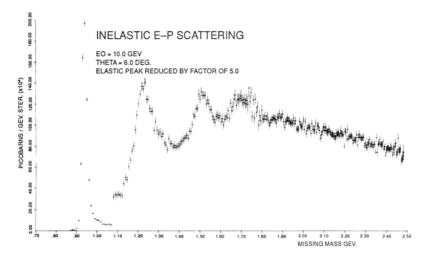

FIGURE 8. *Inelastic electron spectra observed at SLAC with a primary energy of 10 GeV at a scattering angle of 6°.*

one were to add the scattering from all the inelastic channels, the resultant total cross sections would show a fall-off with momentum transfer no more rapid than that expected if the proton were a point. Inelastic scattering on the proton and neutron is a field in its infancy, and it promises to be an exceedingly revealing one.

I have discussed how electron scattering, both elastic and inelastic, can be used to explore the structure of the nucleon, both in its ground state and in excited states. The power of the method rests on the knowledge that the electron is known to interact only through the electromagnetic force, or, in more modern language, through the exchange of a virtual photon, as shown earlier in Figure 2. Let us look at that figure in more detail. Diagrams of this type illustrate the development of elementary particle processes in terms of the intermediate particles (here a photon) that participate in the transition. Frequently, more than one diagram can represent a given reaction, in which case all of them have to be taken into account. Diagrams of this kind have "vertices" at which the lines corresponding to different particles join. At each vertex, the quantum numbers and momenta of the participating particles must balance. However, the *energy* of any intermediate line (called a virtual particle) need *not* agree with the momentum and rest mass of the corresponding free (real) parti-

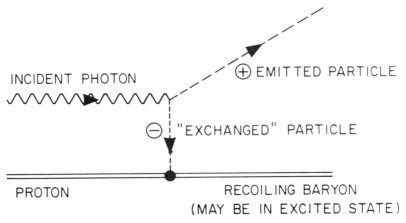

FIGURE 9. *Forward photoproduction of a positive particle. The incident photon produces a "pair" of particles; the positive particle is emitted forward while the negative particle reacts with the proton.*

cle. Since this intermediate particle (the photon, in the case of Figure 2) lasts for a very short time only, its energy cannot be precisely defined, due to the limits set by the uncertainty principle. Such an intermediate or *virtual* particle can thus have an effective mass differing from the *real* particle.

Figure 2 (which shows elastic electron scattering) illustrates exploration of nucleon structure with *virtual* photons. Obviously, we can also study nucleon structure by using *real* photons directly at high energy.

We can use either the free electromagnetic field or the *virtual* electromagnetic field of a scattered electron to generate the new particles of modern physics. The former process is called "photoproduction"; experimentally, it consists of bombarding suitable targets, usually of liquid hydrogen, with high-energy photons in the form of x rays produced by high-energy electron impact and then analyzing the reaction products. A great deal of work has been dedicated to such processes in the past, but some exceedingly exciting results have been obtained lately. I can give you only a very narrow view of this wide field.

Frequently, when a photon is incident on a proton target, one can describe the process by a mechanism of the kind shown in Figure 9. The electromagnetic field of the incident photon creates a new particle that is emitted *forward* and another particle that is then absorbed by the nucleon. If one of these particles is positively charged, the other one must be negative. Similarly, quantum numbers other than electric charge must be con-

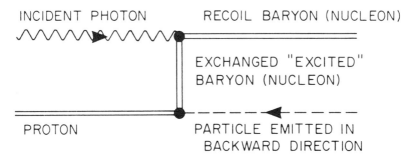

FIGURE 10. *Backward photoproduction. The incident photon interacts with the current of the baryon, which has emitted a new particle in the backward direction.*

served with those of the photon. If one studies the angular and energy distribution of the outgoing particle, one obtains a great deal of insight into the nature of the exchanged particle, which may or may not be a real particle existing in nature. Experiments of this kind have recently been extended into the 20 GeV range of energy; the results have shown that the family of exchanged particles that is required to explain the observed phenomena must be considerably richer than the group of particles that have been discovered in a free state.

Another equally powerful method of examining nucleon structure with photons is to look at production of new particles that are emitted *backward* (Figure 10). The particle exchanged between the incoming photon and the outgoing secondary particle must be an excited state of the nucleon, and one would expect, therefore, that the energy spectrum of the outgoing particle would exhibit bumps similar to those shown before in the inelastic spectra in electron scattering. Figure 11 shows that this is indeed the case; thus we have another new and independent tool for examining the spectroscopy of the nucleons.

Finally, I would like to mention a new and very interesting field in high-energy electron–photon physics. This is the relation of photon physics to some of the most recently discovered particles in physics, the so-called vector mesons. Vector mesons are particles with exceedingly short lifetimes (on the general order of 10^{-23} second) that have the *same* quantum numbers as the photon, which is the carrier of the electromagnetic force and which has zero rest mass; however, the vector mesons have a large rest mass, which, in the case of its most prominent member, the rho meson, is 1500 electron masses. It is therefore possible, in a high-energy

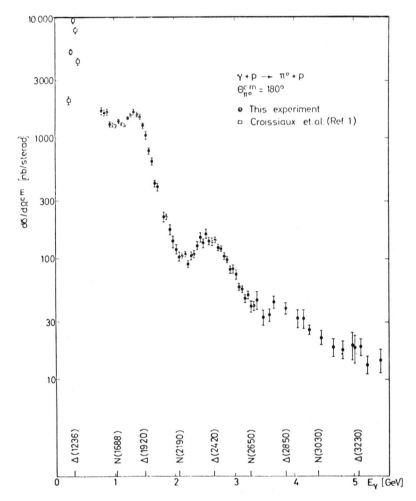

$\gamma + p \longrightarrow \pi^{\circ} + p$
$\Theta^{c\,m}_{\pi^{\circ}} = 180^{\circ}$

● This experiment
□ Croissiaux et.al.(Ref 1)

FIGURE 11. *Backward photoproduction of single neutral pions (DESY data).*

collision, to "create" a vector meson in a collision between a high-energy x-ray photon and a nucleus without exchanging anything but energy, since the quantum numbers of the outgoing vector meson and the incoming photon are the same. Therefore, the nucleus on which this transformation takes place can be left in its ground state, and therefore all the constituents of the atomic nucleus—that is, all the neutrons and protons—can act coherently in producing the reaction (Figure 11). For this reason, the production cross section by photons of such vector mesons may become

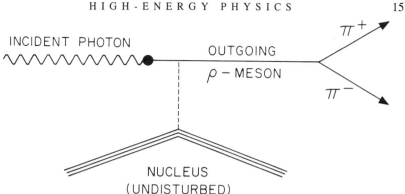

INCIDENT PHOTON

OUTGOING

ρ – MESON

π^+

π^-

NUCLEUS
(UNDISTURBED)

FIGURE 12. *Photoproduction of a vector meson (the ρ meson disintegrates into a pion pair).*

large, if one uses targets of high atomic number. This is indeed the case; Figure 12 shows some of the curves documenting this process. Note that, for very large atomic numbers, the rate of increase of the cross sections turns down again; this is due to the absorption of the rho mesons when traversing the larger nuclei. We can thus derive from these experiments the interaction cross section of the rho meson with nuclear matter—this seems really a remarkable feat if one considers that rho mesons live only for 10^{-23} second.

A great deal can be learned from studying production processes of this kind; however, any process in which a proton is a target and in which a new, unstable particle is formed will necessarily suffer from complexity. The reason is that baryons (that is, protons, neutrons, or hyperons) are conserved in any process; therefore, a proton or another baryon will remain in the final state, and the study of the new, unstable particle will be complicated by its interaction with the proton. It would be very nice indeed if one could invent a means to create these new particles out of pure energy—that is, without the presence of any kind of material target at all. Fortunately, in recent years we have discovered just such a pure process, and this is the use of colliding electron and positron beams. If electrons and positrons collide at low energies, they can annihilate into two photons; however, at high energies they can also annihilate into any other energetically possible combination of final particles that conserves the applicable quantum numbers.

Let me give you an example again in terms of the rho meson, which disintegrates very rapidly into two pi mesons. If an electron and positron collide, a virtual photon can be formed, which in turn can become a rho

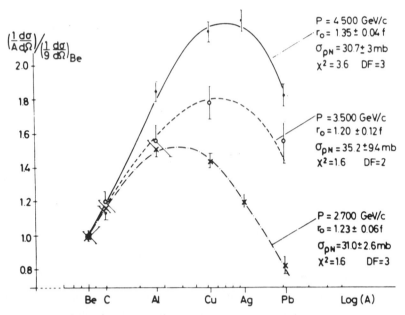

FIGURE 13. *Coherent photoproduction of* ρ *mesons by photons on various elements, shown for various momenta.*

meson, which then can disintegrate into two pi mesons. This chain of events is shown in Figure 13. We therefore have a process in which nothing is left over in the final state except the new, unstable particle, and its properties can be studied without complication from other influences. But how do we do this? How do we produce collisions between electrons and positrons, when positrons do not occur in nature? A first answer might be to build a positron accelerator and let the positron beam hit electrons contained in ordinary atoms. It is indeed possible to accelerate positron beams with an electron accelerator; in fact, the SLAC 2-mile accelerator does produce a very intense beam of positrons up to an energy of about 12 GeV. However, the difficulty is that if a beam of electrons of energy E hits a stationary electron of mass m, then, according to the special theory of relativity, the energy available in the reaction is given by $\sqrt{2\,Emc^2}$. This quantity is generally small, since most of the energy of the electron is used up to set the common center of mass of the particles into motion and very little of it goes into the energy of the interaction. Numerically, the positron beam of 12 GeV, such as the one from the SLAC accelerator,

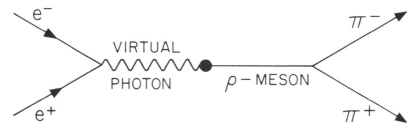

FIGURE 14. *Production of a ρ meson in electron–positron collisions. Note the absence of nucleons in process.*

would produce a reaction energy of only 110 MeV, which is inadequate for studying the kinds of processes of interest and which would be insufficient to create a particle like the rho meson. The situation would be vastly improved if one were to use two accelerators, one accelerating electrons and the other positrons, and bring the two beams into collision. In such a case, the center of mass remains at rest and the sum of the energies of the two beams is entirely available for the reaction. This is the basic idea underlying "colliding beam" accelerators. Energetically, this argument appears to be all right, but, with respect to intensity, we are dealing with a serious problem, since the density of a practical electron beam is lower than the density one is able to obtain in a good laboratory vacuum. Therefore, collisions between two external linear beams would not produce a sufficient reaction rate. However, this problem can be solved by storing both beams in what is known as *storage rings*—that is, annular magnetic fields that can confine the beams. In that case, the two beams can cross one another repeatedly and the resulting increase in reaction rate, combined with the simplicity of the reaction and the high reaction energy, makes such storage-ring, colliding-beam devices very attractive indeed. Colliding-beam installations at which important physics experiments were first done were built in the United States; in recent times, however, initiative in this field has shifted to Western Europe and the Soviet Union.

Figure 15 shows such an installation, located at Novosibirsk, which stores negative electrons and positrons at an energy of 750 MeV. Figure 16 shows one of the results obtained with this device. The measurement determined the rate of a reaction discussed earlier—namely, the formation of a rho meson and its subsequent disintegration into pi mesons resulting from electron–positron collisions. The curve shows this rate as a function

FIGURE 15. *Storage-ring and colliding-beam installation (VEPP-2) at Novosibirsk, USSR. Electrons and positrons, each of 750 MeV, can be stored here.*

of the energy of the particles stored in the two rings. You can see that a very pure resonance curve results, from which both the energy and the width of the rho state can be deduced. Interestingly enough, these numbers differ from those inferred from experiments on the rho meson with conventional accelerators, indicating that interactions of the final rho meson with the unavoidable other final products had distorted the earlier results. Colliding-beam accelerators are thus indeed a uniquely valuable addition to electron and photon physics, and they probably constitute the ultimate tool for "exploring" the unknown particles with the known forces of electrodynamics because they permit the greatest possible isolation of the unknown object.

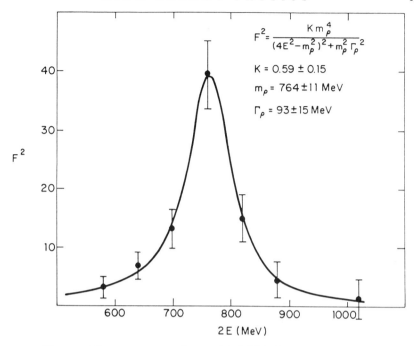

FIGURE 16. ρ *resonance observed in VEPP-2 at Novosibirsk.*

Thus far, I have strictly adhered to my topic: "High-Energy Electron and Photon Physics: Exploring the Unknown with the Known," in which by "the known" I mean quantum electrodynamics. But is this basic premise really true? Do we know that quantum electrodynamics remains valid at the exceedingly high energies and small distances that constitute the sphere of interest of high-energy physics? As mentioned before, quantum electrodynamics has been verified quantitatively over an unprecedentedly large range of energies and distances, ranging to energies in the GeV region and to distances down to 10^{-15} centimeter. Naturally, there is a great deal of interest to push these boundaries still further—that is, to explore whether quantum electrodynamics is valid at even smaller distances. To put it into more elementary terms: We wish to explore whether Coulomb's law, which implies that the electrostatic potential would vary inversely with distance from a point source, would still continue to increase at these exceedingly small distances. Beyond natural curiosity and our desire to extend our knowledge of the range of validity of any physical law, including quantum electrodynamics, there are also more specific incentives for examining this question. The most important of these remains the fact that no convergent theory of that part of the mass of elementary particles can be made that originates from electromagnetic energy unless one assumes in some way that Coulomb's law ceases to be valid at very small distances. It may, of course, be that posing the question of the origin of an electromagnetic mass is not meaningful, but the existence of the question gives an additional impetus to examine the range of validity of electrodynamics at small distances and, therefore, at high-momentum transfer. Unfortunately, the experiments required to examine these questions are difficult to design and to execute. For one thing, such experiments must be designed so that the influence of an unknown or poorly known structure (such as that of the proton or of other hadrons) on the desired result will be avoided. Experiments to examine the validity of quantum electrodynamics are therefore those that either involve only photons, electrons, or muons or those that involve protons only in such a way that the effect of proton structure can be either eliminated or neglected.

In addition, if we believe in the basic assumptions of special relativity, then the momentum that defines the scale of the process must be the so-called four-dimensional momentum q; the square of its magnitude is given by $q^2 = E^2 - c^2p^2$, in which E^2 is the energy involved in a given process and p^2 is the ordinary, three-dimensional momentum. It can be shown that this quantity is always the square of the rest mass of each particle; this is 0 for the free photon, and it is small for such particles as the free electron and muon, which do not carry nuclear interactions but carry the electromagnetic interaction only. However, if we invoke processes in which these particles are virtual rather than real, then this critical quantity may become large. Thus, the kinematical conditions must be such that electrons, muons, or photons are in a virtual state having a large "four-momentum" q.

Time does not permit me to give an exhaustive summary of the experiments that fulfill these various requirements. The most important ones have been the study of the production of electron–positron pairs at large angles of production by high-energy incident photons, study of the gyromagnetic ratio of the muon, and study of electron–electron collisions in colliding-beam experiments. The first experiment, wide-angle electron–positron pair production, has been investigated at laboratories in the United States and Western Europe. For some time, one of these experiments seemed to give clear indication of a violation of the laws of quan-

FIGURE 17. *The Stanford–Princeton colliding-beam installation. Electrons with an energy of 550 MeV are stored in each of the two rings and collide in the common section.*

tum electrodynamics, but more recent and careful determinations indicate complete agreement. These studies of the wide-angle electron–positron pair production process form one of our principal anchor points at this time, giving the smallest distance over which the laws of electrodynamics seem to remain valid. Our most stringent limit on this question, derived from a purely electrodynamic process, comes again from storage-ring studies. Figure 17 shows an experimental configuration, originating from a collaboration between Princeton University and Stanford University, in which electrons are stored in magnetic rings that have a common section. Electrons are stored in each ring at an energy of 330 MeV and they make collisions in the common section. The collision cross section is studied as a function of the angle between the emerging electrons and the incident collision path. Figure 18 shows the agreement between the observed data and the predictions from calculations of quantum electrodynamics. As can be seen by inspection, and as can be verified by more quantitative statistical analysis, agreement between experiment and theory is excellent. We thus conclude from these and other experiments that, at this time, quan-

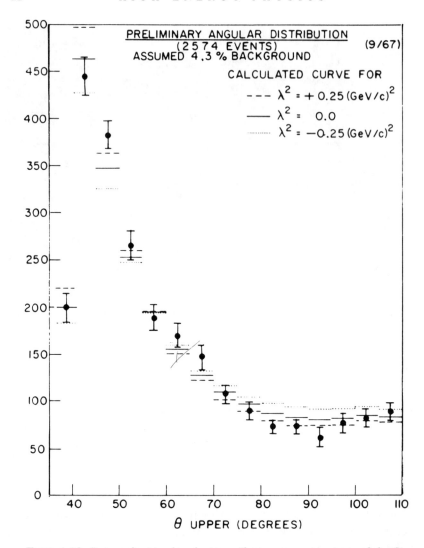

FIGURE 18. *Data taken in the electron–electron storage rings of the Stanford–Princeton installation shown in Figure 17. The experimental points are the observations on elastic electron–electron scattering as a function of angle. The solid horizontal bars are the predictions from quantum electrodynamics. The dashed lines correspond to breakdown models of quantum electrodynamics at distances on the order of 10^{-14} centimeter.*

tum electrodynamics remains a quantitative theory whose validity extends all the way from the largest macroscopic distances to subatomic distances of a small multiple of 10^{-15} centimeter. Thus, high-energy electron and photon beams remain a tool with which we can explore the unknown structure of the new subatomic particles with known forces.

Colliding Beams versus Beams on Stationary Targets: Competing Tools for Elementary-Particle Physics

H igh-energy accelerators have been the primary tools in advancing elementary particle physics ever since J. J. Thompson discovered the electron, using what, in modern terms, might well be called the first accelerator. Although many of the qualitative findings, including the discovery of new unstable particles, originated from cosmic-ray studies, quantitative measurements have required accelerators as particle sources. In fact, as the energies of accelerators have pushed to higher and higher values, the relative importance of cosmic rays as a tool for the study of any aspect of elementary particle physics has sharply diminished, although study of cosmic-ray physics remains an important activity in its own right as a diagnostic tool of the cosmos.

Figure 1 shows the state of the accelerators in the world in an admittedly oversimplified manner by plotting their energy and intensity: the power of accelerators cannot be measured by one parameter—or even by two parameters. Particle energy remains, of course, the foremost quantity of interest, but intensity, beam quality and geometry, and many other factors determine the usefulness of these instruments to the experimental physicist. The chart shows that the National Accelerator Laboratory cur-

A talk given at National Academy of Sciences, October 16, 1972.

rently holds the world's record in energy—300 giga-electron-volts (300 GeV)—and aspires to further increasing values; the collaborative international laboratory at Geneva (CERN) has an accelerator under construction aiming at similar energies. These two machines are proton machines; the energy record for electrons is held by SLAC (22 GeV) with the aspiration of expanding the energy to about 50 GeV with the aid of a recirculation scheme.

Elementary-particle physics has been exceedingly productive throughout this century, and it is fair to say that the time interval between new discoveries in this field, which have changed our basic view of nature, has not shown any signs of stretching out. Yet the magnitude of the tools and the concomitant cost of operating them have grown steadily, and the question is being raised with increasing frequency as to when and how this evolution might stop. As shown in Figure 2, the trend in the increase of accelerator energy has not stopped. New inventions have sustained an almost exponential increase in energy of one decade every six years, even though the scaling laws pertaining to each type of accelerator have in the past forced a leveling off of the energy attainable by any one method. This increase in energy has been bought at a serious social cost: Because of the high price of each accelerator, the total number of installations worldwide that operate at the frontiers of the field has been steadily decreasing. Therefore, elementary-particle physicists at academic and other institutions have had to carry out their experimental observations away from home. Yet the consensus remains that involvement of academic physicists within the field should be maintained or even strengthened despite this difficulty: Elementary-particle physics, since it is among the most basic of the sciences, remains an essential ingredient of the educational program in physics at major universities.

A possible departure from the pattern outlined above has been introduced by the emergence of colliding-beam techniques. It has been recognized for a very long time that the threshold for a reaction among particles to occur is set by the "center-of-mass energy"—that is, by the energy measured in that frame in which the center of mass of the colliding system is at rest. At highly relativistic particle energies, the center-of-mass energy increases only with the square root of the energy (as measured in the laboratory) of the particle that bombards a stationary target; the rest of the energy is converted into the kinetic energy of motion of the center of mass of the combined system. The relationship between the center-of-mass energy and the laboratory energy of particle beams striking station-

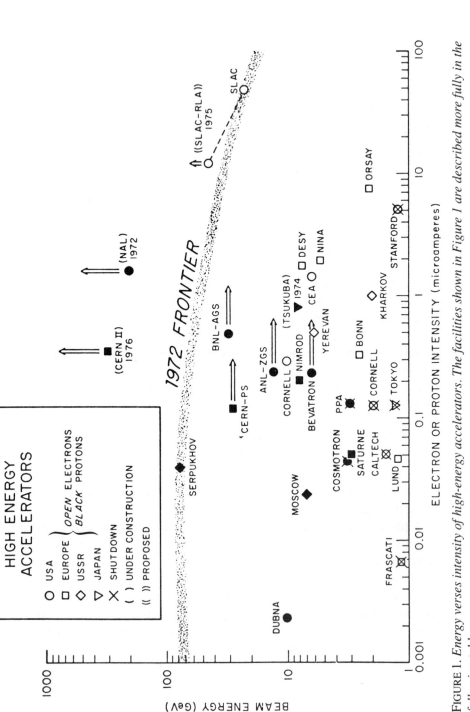

FIGURE 1. *Energy verses intensity of high-energy accelerators. The facilities shown in Figure 1 are described more fully in the following table.*

Machine	Energy (GeV)	Figure 1 name	Institution	Location	Remarks
	1	Frascati	Laboratori Nazionali del CNEN	Rome, Italy	Shut down
	1	Lund	University of Lund	Lund, Sweden	
	1	Tokyo	Tokyo University	Tokyo, Japan	Shut down
	1.5	Caltech	California Institute of Technology	Pasadena, California	Shut down
	2	Cornell	Cornell University	Ithaca, New York	Shut down
Electron Synchrotron	2.5	Bonn	University of Bonn	Bonn, W. Germany	
	4	NINA	Daresbury Nuclear Physics Lab.	Daresbury, U.K.	
	6	Yerevan	Yerevan Physical Institute	Yerevan, USSR	
	6	CEA	Cambridge Electon Accelerator	Boston, Massachusetts	
	7	Desy	Deutches Elektronen-Synchrotron	Hamburg, W. Germany	
	10	Cornell	Cornell University	Ithaca, New York	
	1	Stanford	High-Energy Physics Lab., Stanford University	Stanford, California	Shut down
Electron Linac	2	Kharkov	Physico-Technical Institute	Kharkov, USSR	
	2	Orsay	University of Paris	Orsay, France	
	22	SLAC	SLAC, Stanford University	Stanford, California	
	50	SLAC–RLA	SLAC, Stanford University	Stanford, California	Under design

FIGURE 1 (cont.). *Listing of facilities shown in illustration. Figure continues on next page.*

Machine	Energy (GeV)	Figure 1 name	Institution	Location	Remarks
	3	Cosmotron	Brookhaven National Laboratory	Upton, New York	Shut down
	3	Saturne	Centre d'Etudes Nuclieres	Saclay, France	
	3	PPA	Princeton–Pennsylvania Accelerator	Princeton, New Jersey	Shut down
	6	Bevatron	LBL, University of California, Berkeley	Berkeley, California	
	7	Moscow	Institute of Experimental and Theoretical Physics	Moscow, USSR	
	7	Nimrod	Rutherford Laboratory	Harwell, U.K.	
Protron Synchrotron	8	Tsukuba	National Laboratory for High-Energy Physics	Tsukuba, Japan	Ready in 1974
	10	Dubna	Joint Institute for Nuclear Research	Dubna, USSR	
	12	ANL-ZGS	Argonne National Laboratory	Chicago, Illinois	
	28	CERN-PS	European Organization for Nuclear Research	Geneva, Switzerland	
	30	BNL-AGS	Brookhaven National Laboratory	Upton, New York	
	76	Serpukhov	Institute for High Energy Physics	Serpukhov, USSR	
	200	NAL	National Accelerator Laboratory	Batavia, Illinois	Ready in 1972
	300	CERN II	European Organization for Nuclear Research	Geneva, Switzerland	Ready in 1976

FIGURE 1 (cont.). *Listing of facilities shown in illustration.*

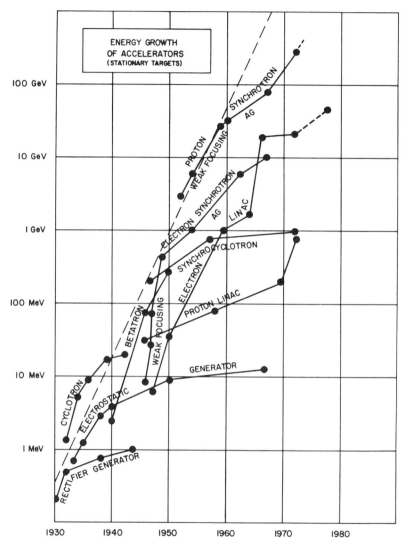

FIGURE 2. *The energy growth of accelerators from 1930 to the present.*

ary targets is shown in Figure 3. Clearly, this decreasing efficiency, in terms of center-of-mass energy, could be circumvented by two beams colliding with each other from opposite directions. This idea is an old one; it is, in fact, difficult to document how it originated. However, the question here is not that of the idea but of its execution. The problem is that the

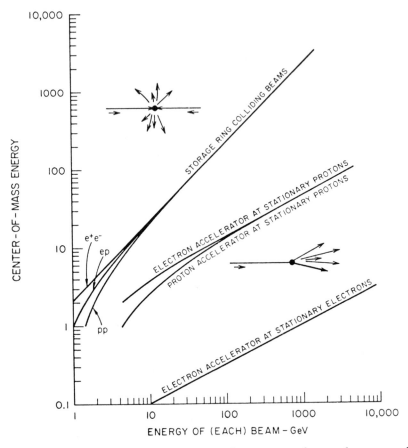

FIGURE 3. *Center-of-mass energy versus beam energy for accelerators and storage rings.*

density of particle beams is vastly inferior to that of ordinary condensed matter and is, in fact, comparable to that of practically attainable vacua; thus, the reaction rates in colliding-beam experiments are apt to be very much lower than those encountered when particle beams strike stationary targets. Quantitatively, this factor is measured by a quantity that colliding-beam physicists call the "luminosity." This is the number by which one multiplies the cross section of the reaction under investigation in order to obtain the reaction rate. Luminosity is therefore generally measured in units of cm^{-2} s^{-1}.

During the last decade, there have been many developments that have demonstrated that the luminosities of colliding-beam devices can reach a range practical for important experiments in elementary-particle physics. The first such demonstration was made in the Stanford–Princeton collaboration, in which two electron beams, each up to 550 MeV, were made to collide in the common section between two magnetic storage rings arranged in a figure-eight pattern. This installation resulted in a pioneering demonstration that quantum electrodynamics remained valid up to energies that had previously been unattainable. The most important and also most ambitious single step in colliding-beam technology was taken in Europe at CERN in their ISR (Intersecting Storage Ring) project, which became operational in 1970. In this installation, particles are injected into two rings at an energy of 26 GeV from CERN's proton synchrotron; the energy in each ring can be slightly raised above that value by increasing the magnetic field. When two 30 GeV beams collide in one of the eight interaction regions of that machine (interaction regions are formed by intersections between straight sections of each ring), the energy in the center of mass of the colliding protons (60 GeV) is equivalent to that produced by a beam of 1900 GeV striking a stationary target. The luminosity of this installation has reached a figure somewhat below 10^{30} cm^{-2} s^{-1}. Since the total cross section for reactions at very high energies between colliding protons is somewhere near 10^{-25} centimeter squared, the total interaction rate is in the neighborhood of 10^5 reactions per second. This figure in itself is, of course, not fully persuasive as to the usefulness of the installation, since not all the reactions produced would be of equivalent interest; for instance, reactions incurring large momentum transfer are, in general, the most sensitive probe of unknown features of strongly interacting particles, but the cross sections decrease steeply with an increase in this quantity. Practical limits set by the art of particle detection would also reduce the number of observed events per second below this figure. Nevertheless, the results obtained with the ISR installation thus far have amply demonstrated the exciting possibilities of colliding-beam techniques. Many experiments reported from the ISR have yielded important new knowledge on the phenomenology of high-energy reactions and on the behavior of cross sections governing pairs of particles at these exceedingly high energies. Moreover, searches for new particles and for qualitatively new phenomena carried out in this installation have set an upper limit on what might be observable in the future.

Several colliding-beam installations other than the original Princeton–Stanford storage rings and the powerful ISR have also been built. Figure

FIGURE 4. *Aerial view of the Stanford Linear Accelerator Center, showing the recently completed storage ring, SPEAR.*

4 tabulates the status of these storage-ring facilities worldwide. Note that the CERN ISR and the SLAC installation SPEAR hold the current records in energy and luminosity for protons and electrons, respectively; the future evolution of colliding-beam devices will be discussed shortly.

There are, of course, many differences, other than center-of-mass energy and attainable data rate, that distinguish the kinds of physics that can be done with conventional accelerators producing beams striking stationary targets from those possible at colliding-beam installations. Conventional accelerators with stationary targets not only permit the study of primary interactions between the beam particles and the constituents of the target but they also function as "factories" of secondary beams of unstable particles. Such secondary beams are frequently at least as valuable as the primary beam in studying elementary-particle interactions of a well-identified nature. The availability of secondary beams means that a "conventional" accelerator can service a very large number of experimental stations and thus support a larger community of particle-physics experiments; in contrast, the number of experiments at a colliding-beam facility is generally restricted to the number of "interaction regions."

The use of conventional accelerators in which either primary or secondary beams strike material targets (usually liquid hydrogen) introduces a complicating factor: Since baryons (protons and neutrons, or one of their "strange" partners, such as the lambda particle) are conserved in any elementary-particle process, the struck proton in the collision either will be present in the final state or will have changed to another baryon. Accordingly, in reactions in which, for instance, a pair of pions of opposite charge is created in the collision, a nucleon will also be present; therefore, the final state is a three-body rather than a two-body system. Thus the interaction between two pions in isolation cannot be studied with a conventional accelerator. In contrast, when, for instance, electrons and positrons collide in a storage-ring arrangement, these two particles annihilate into purely electromagnetic energy known as a "virtual photon." This virtual photon is at rest in the laboratory and can rematerialize into any combination of particles in the final state that obey the conservation laws applicable to that situation. Specifically, the virtual photon is neutral, has spin 1, and has negative intrinsic parity. According to the conservation laws, this permits, for example, a pair of positive and negative pions or kaons to be formed in the final state, or one of the so-called vector mesons to be created (these generally decay into pairs of pions or kaons). These objects are thus made available in the laboratory without the disturbing influence of a baryon, so the interaction between pairs of unstable particles can be studied under conditions of much greater simplicity than is possible with conventional accelerators. Not only is the absence of a baryon a simplifying factor, but also the well-defined quantum numbers of the initial state simplify analysis of the unknown final-state interaction, since the number of spectroscopic "states" in the final state is constrained.

"Conventional" accelerators and storage rings need not be separate installations; on the contrary, a conventional accelerator can serve the dual purpose of injecting particles into a storage ring and of supporting a research program in its own right. This is the case at SLAC, which is shown in Figure 5. Here the 2-mile accelerator operates both at the energy and intensity frontier of electron machines, but it also injects electrons and positrons into a colliding-beam storage ring called SPEAR; this device operates at present also at the highest luminosity and energy of existing electron colliding-beam installations. Similarly, the CERN international

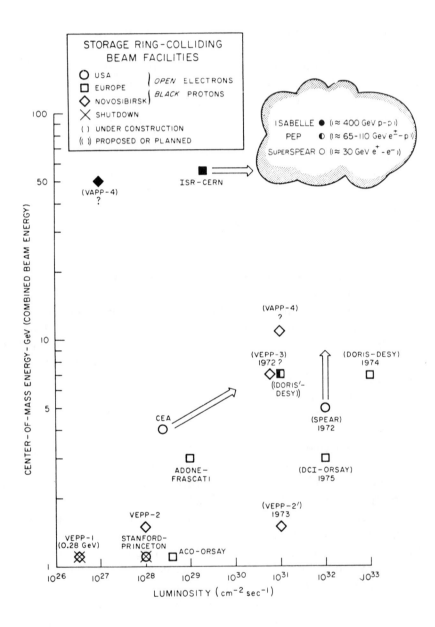

FIGURE 5. *Center-of-mass energy versus luminosity for storage-ring collid-ing-beam facilities. The facilities shown in Figure 5 are described more fully in the following table.*

Particles stored	Energy of each beam (GeV)	Fig. 5 name	Institution	Location	Remarks
Electrons + Electrons	0.14	VEPP-1	Institute of Nuclear Physics	Novosibirsk, USSR	Shut down
	0.55	Stanford–Princeton	High-Energy Physics Laboratory, Stanford University	Stanford, California	Shut down
Electrons + Positrons	0.55	ACO-Orsay	University of Paris	Orsay, France	
	0.75	VEPP-2	Institute of Nuclear Physics	Novosibirsk, USSR	
	0.75	VEPP-2'	Institute of Nuclear Physics	Novosibirsk, USSR	Ready in 1973
	1.5	Adone-Frascati	Laboratori Nazionali del CNEN	Rome, Italy	
	1.5	DCI-Orsay	University of Paris	Orsay, France	Ready in 1975
	2.5	SPEAR	SLAC, Stanford University	Stanford, California	Operating in 1972
	3.5	CEA	Cambridge Electron Accelerator	Boston, Mass.	
	3.5	VEPP-3	Institute of Nuclear Physics	Novosibirsk, USSR	Ready in 1972?
	3.5	Doris-DESY	Deutsches Elektronen Synchrotron	Hamburg, Germany	Ready in 1974
	6	VAPP-4	Institute of Nuclear Physics	Novosibirsk, USSR	Date uncertain
	15	SuperSPEAR	SLAC, Stanford University	Stanford, California	Under study
Electrons + Protons	3.5	Doris-DESY	Deutsches Elektronen Synchrotron	Hamburg, Germany	Under study
	15 e^- or e^+ / 70 or 200 p	PEP	SLAC, Stanford + LBL, Berkeley	Stanford, Berkeley, California	Under study
Protons + Protons	28	ISR-CERN	European Organization for Nuclear Research	Geneva, Switzerland	
	200	ISABELLE	Brookhaven National Laboratory and Collaborative	Upton, New York	Under study
Protons + Antiprotons	24	VAPP-4	Institute of Nuclear Research	Novosibirsk, USSR	Date uncertain

laboratory at Geneva operates the ISR and its injecting proton synchrotron as a system of separate research tools.

To summarize, we find that conventional accelerators producing beams striking stationary targets and storage rings producing colliding beams have both assets and deficiencies for high-energy particle physics. Generally, the conventional accelerators yield much higher intensities and produce useful secondary beams, whereas storage rings can produce much higher center-of-mass energies and can produce states of interaction that are easier to analyze in terms of fundamental questions. Thus normally one would conclude that both of these accelerator types have their proper area of usefulness and that both should be developed further. This statement implies that there are no technological or financial limits toward pursuing either or both directions; in practice, of course, both of these limits are very real. Technology alone at this time does not impose any substantial basic upper boundary on the performance of conventional accelerators. The "alternating gradient synchrotron" using conventional magnets can, in principle, be extended to any arbitrary *energy*, provided that limits of real estate or the taxpayers' willingness do not intervene. Once development of reliable superconducting magnets has been completed, their substitution would reduce the required space, but not necessarily the cost, of an accelerator installation. There are technological limits governing the *intensity* attainable by proton synchrotrons and electron linear accelerators; these derive in part from the characteristics of practical injection systems, in part from radioactivity and the problem of designing targets for very intense beams, and also in part from certain factors set by beam-orbit dynamics, which limit the intensity of the beams that can be accelerated. Ultimately, however, both the energy and the intensity limits of conventional accelerators appear to be financial rather than physical.

The intensity limitations on storage rings, in contrast, are more fundamental. In a storage ring, the two beams that are destined to collide in the interaction regions must be stored in a magnetic guide field for a period of time measured in minutes or hours. This poses extreme requirements on vacuum technology; moreover, in the case of electrons, energy is radiated during the storage process, requiring very high radiofrequency power to compensate for this loss. However, most basic is the fact that such stored beams must be *stable*, and it has been the history of each new generation of storage rings that new sources of instability have been discovered whenever a new design entered the test phase. There are many such instabilities: both those associated with the storage of a single beam and those associated with the interaction between the colliding beams. Single-beam instabilities can, in principle, be controlled by feedback mechanisms, although this can be exceedingly difficult in practice. The interaction between the beams produces modifications of the focusing conditions confining each beam; if such shifts become too large, this produces instabilities that are in principle difficult, if not impossible, to prevent. Such beam–beam instabilities thus set a limit to the practically attainable luminosity of storage rings. Several inventions have been made in recent times that advance this limit substantially, but, even so, it is clear that the interaction rates of storage-ring–colliding-beam installations will always be many orders of magnitude below those attainable in interactions of the *primary* beam found in conventional accelerators with a stationary target. However, the interaction rates become quite comparable if colliding-beam installations are compared with the use of *secondary* beams from conventional accelerators.

Figure 6 attempts to make a more quantitative comparison between the performance of storage rings and that of conventional accelerators, albeit in a highly oversimplified way. The chart plots the "effective luminosity" of the installation in question against the center-of-mass energy. The luminosity of a conventional accelerator depends, of course, on the thickness of the target actually employed, and we assume in this chart that such a target is a 1-meter-long vessel of liquid hydrogen. The chart also assumes that the efficiency for detecting the products of reactions produced both in colliding-beam devices and in conventional accelerators is 100 percent—that is, that all the reaction products are seen. Note that the chart spans a range of approximately 20 decades in luminosity, while the installations shown in the chart range some two decades in center-of-mass energy (which would be *four* decades in laboratory energy for conventional accelerators).

The chart shows, inside a dashed rectangle, a series of installations that are at present under active discussion for the future. Such studies are being carried out both on the West Coast, with emphasis both on electron–positron and on lepton–proton storage rings, and on the East Coast, with emphasis on proton–proton storage rings; similar studies are in progress in Europe. These studies project the storage-ring art for protons and electrons well beyond that available at the ISR at CERN and the SPEAR electron–positron ring at SLAC. These studies have been encouraged by the very high productivity experienced with the ISR installation and the suc-

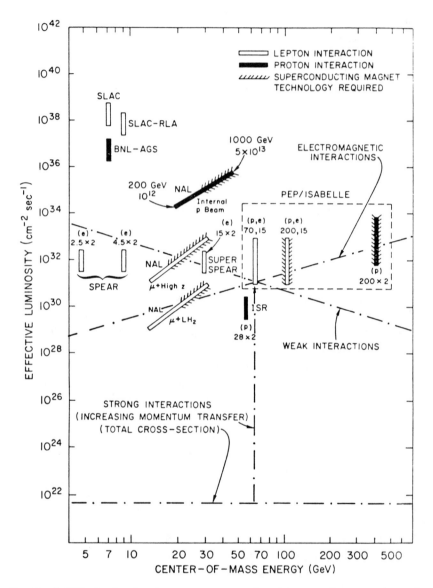

FIGURE 6. *Effective luminosity versus center-of-mass energy for several accelerators and storage rings, existing and under study. The facilities shown are identified in previous figures, except for the following:*
(a) SuperSPEAR is a study being carried out at SLAC of the possible characteristics and uses of a colliding-beam storage ring that would store beams of electrons and positrons up to energies of 15 GeV (each beam).
(b) PEP is a study being carried out by a collaborative group from SLAC and

LBL, Berkeley, on the possible characteristics and uses of a colliding-beam storage ring that would permit collisions between 15 GeV electrons and 15 GeV positrons, or between either of these particles and protons with energies between 70 GeV and 200 GeV.

(c) ISABELLE is a study being carried out by a group from Brookhaven National Laboratory and collaborators on the possible characteristics and uses of a colliding-beam storage ring that would permit collisions between beams of protons having energies up to 200 GeV (each beam).

Figure 6 attempts to display both accelerator and storage-ring installations on a comparable scale. Naturally, such an attempt will involve some oversimplification. The data rates attainable are described by an "effective luminosity"; this is, the number by which the cross section (measured in square centimeters) of the reaction channel under observation is to be multiplied to arrive at a rate in events per second. This scale replaces the "intensity" figures, in microamperes or in particles per pulse, that are usually displayed for accelerators. It is assumed that the reaction in question is observed at 100 percent efficiency, and that the detector solid angle collects all the events of interest. To apply this concept to an accelerator, it is assumed that (unless otherwise indicated) a liquid-hydrogen target 1 meter in length is used. With the exception of the muon-beam entries, all figures refer to primary beams.

Center-of-mass energies are plotted under the assumption of a stationary proton target, in the case of conventional accelerators. Those U.S. accelerators that are operating or are under study, and that have a center-of-mass energy greater than 5 GeV, are listed in the figure. The CERN ISR is shown for comparison with the U.S. colliding-beam storage-ring projects under consideration. CERN II is not explicitly shown, but its performance would be comparable to NAL. NAL performance is shown under a wide range of assumptions; these range from an energy of 200 GeV at 10^{12} protons per pulse (which might be available for physics research by the end of this calendar year) all the way up to an intensity of 5×10^{13} protons per pulse at an energy of 1000 GeV. The latter values are very optimistic assumptions, in regard both to intensity and to the feasibility of the superconducting "doubler" project for NAL.

Since the engineering feasibility of large-scale superconducting-magnet technology has not been demonstrated, a special notation is made in the figure to point out those devices that would require such technology.

The dash–dot lines on the figure indicate the luminosities that are required to achieve a counting rate of one event per hour, at the center-of-mass energies shown, for weak, strong, and electromagnetic interactions. The vertical dash–dot line extending upward from the strong-interaction line is meant to point out the increasing luminosities needed for rates of one count per hour for events of increasing momentum transfer.

cessful initial turn-on of SPEAR. You will note that the center-of-mass energy in the most ambitious of these studies might reach as high as 400 GeV; this is the same center-of-mass energy as that obtained when a beam of protons with an energy of 80,000 GeV strikes a stationary target, clearly an impossible goal using ordinary means of acceleration. In looking to the future, the primary problem is therefore whether such very large storage rings can in fact be built *at a useful luminosity*, and from this arises the question, "What is a useful luminosity?"

The question of minimum usable luminosity is, of course, associated with the projected areas of interest on which high-energy physics might focus in this ultrahigh-energy region. Adequate discussion of this topic would lead much too far afield. However, for illustration, let me indicate regions in Figure 6, which shows the minimum luminosity required to produce a counting rate of one count per hour for processes driven by the three dominant forms of interaction in elementary-particle physics: the *electromagnetic* interaction, the *weak* interaction, and the interaction among *strongly interacting* particles known as hadrons. These regions are shown in Figure 6. The cross section for the electromagnetic interaction between two charged particles that leads to pointlike end products is expected to *decrease* inversely as the square of the center-of-mass energy; the validity of the theory that predicts this behavior has been demonstrated thus far over the full range of energy studied to date. Moreover, studies of the reaction

$$e^+ + e^- \rightarrow \text{any combination of hadrons}$$

over the range of energies accessible to electron–positron storage rings to date have shown that the cross section for this process decreases with energy no faster than that of the pointlike process (such as $e+ + e^- \rightarrow \mu^+ + \mu^-$), and in fact, exceeds it in magnitude.

The weak interaction is expected to exhibit a cross section that *increases* linearly with the square of the center-of-mass energy, and Figure 6 indicates that, at an energy near 100 GeV, the weak and the electromagnetic cross sections might cross over.[2] It is, of course, generally recognized that no cross section in nature can increase indefinitely with energy, and that something has to "go wrong" with the theory before the so-called

---

[1] The reference electromagnetic cross section used in Figure 6 is that of the reaction $e^+ + e^- \rightarrow \mu^+ + \mu^-$.

[2] The reference weak-interaction cross section used in Figure 6 is that of the reaction $e^- + p \rightarrow$ neutrino + any combination of hadrons.

"unitarity limit" for quantum mechanics is reached. Therefore, particle physicists confidently expect that the Fermi theory of weak interactions, in which four spin 1/2 particles are assumed to interact at a mathematical point, will not remain valid to indefinitely high energies; eventually this interaction will have to be carried by some kind of mediating particle.

The total cross section derived from strong interactions is expected to remain roughly constant, as is also shown in Figure 6.[3] However, the partial cross sections of greatest interest are much smaller: these involve large transfers of momentum among the colliding particles; Figure 6 shows that partial cross sections of as low as 10^{-10} times the magnitude of the total cross section might be accessible to measurement at storage rings whose luminosities are adequate for the observation of the weak and electromagnetic interactions. One might speculate that, at corresponding momentum transfers, the strong interaction cross sections approach those of weak and electromagnetic interactions at energies in the center of mass near 100 GeV.

It appears plausible, therefore, that near a center-of-mass energy near 100 GeV (which is an energy nobody dreamed of reaching in the laboratory until a relatively short time ago), the basic distinction between weak, strong, and electromagnetic interactions might lose meaning, at least as far as signifying magnitudes of cross sections is concerned. There are some members of the theoretical community who are optimistic that work at these extremely high energies will lead to the discovery of unifying principles that will combine their understanding of these different forms of interactions, which have thus far been treated in an essentially separate manner; unification of weak and electromagnetic interactions has already been attempted with some success. Whether this highly ambitious goal will in fact be reached with such installations is, of course, for the future to decide; even from a more simplistic point of view, it is clear, however, that storage rings permit, in principle, an enormous extension of the kinematic parameters accessible to experimenters that is simply not possible by any other means.

Figure 6 should, however, induce some caution; in the "exciting" region at which these curves converge, a luminosity of 10^{32} cm^{-2} s^{-1} would yield only a few counts per hour, and this value should thus be considered a practical minimum objective for storage-ring installations. Orbit theo-

---

[3] The reference "strong" cross section used is the total cross section for the reaction $p + p \rightarrow$ anything.

rists are optimistic that such luminosity values can be obtained; the SPEAR storage ring at SLAC is now only one order of magnitude short of this goal, while the ISR operates currently two orders of magnitude below this value.

Some rough economic comparisons might also be useful. Without going into detail, it appears that the generation of super storage rings will cost less than the NAL accelerator, now in its final test phase near Chicago, or the CERN-II accelerator, now under construction at Geneva. It is certain that the cost will be about an order of magnitude below that of any follow-on conventional accelerator installation exceeding 1000 GeV in energy.

Let me summarize: The evolution of sources of high-energy particles for use in elementary-particle physics is now approaching a fork in the road. One path leads to "conventional" accelerators of higher and higher energy, and the other leads to "super storage rings," which promise attainment of center-of-mass energies and access to new physics, hitherto unimaginable, but which require advances in luminosity in order to be practically useful. Considering these facts and the expectation that the second path promises advances at lower cost, it will be pursued in the future with increasing enthusiasm. Whether another step along the former path will be taken, either in the United States or abroad, is a matter of intensive current discussion, in particular in the Soviet Union. I hope, however, that I have presented an overview that has been persuasive in showing that the field of elementary particles is very far from being "run dry," either in terms of expected fundamental findings or in terms of available technology to provide the tools.

Special Relativity Theory in Engineering

The title of this talk may, in itself, be a surprise to some, in that special relativity is frequently portrayed as being mysterious, abstract, and in no way related to the technological world. This is far from the truth. I will talk today about the role that special relativity plays in engineering applications. In other words, I will talk about certain kinds of apparatus that perform useful (or useless) functions and in which special relativity plays a necessary role in their design. In fact, such devices simply would not work if special relativity were incorrect. In this sense, such devices act as a complement to experiments that are especially designed to test the validity of special relativity.

I will discuss here the intellectual content of special relativity only very briefly. Special relativity resulted from a body of experimental knowledge and theoretical conjecture that accumulated over the last quarter of the previous century. Mechanics had become an exact science, and its laws exhibited the principle, known as Galilean relativity, that identical experiments that were carried out in laboratories, moving uniformly with respect to one another, should not exhibit different results. At the same time, Maxwell's equations had provided a full understanding of phenomena involving electricity and magnetism, including the propagation of electromagnetic radiation. Among other consequences, these equations imply that, in a vacuum, the velocity of propagation of such radiation—that is, the velocity of light—is a constant.

A lecture given at Institute for Advanced Study Symposium, "Einstein Centennial," March 5, 1979.

A large amount of observational material relating to light emission from moving sources, to light propagation in moving media, and to moving electromagnetic devices had led to a series of paradoxes that appeared to indicate a conflict between Maxwell's equations, on the one hand, and the principle of Galilean relativity, on the other. Einstein proposed that all of these paradoxes could be resolved by modifying the laws of mechanics rather than those of electromagnetism. Specifically, he postulated that the mathematical relations that described the dependence of electromagnetic phenomena on the states of motion of the observer and the physical phenomena should also apply to the laws of mechanics. In so doing, the velocity of light, which previously had not played a special role in mechanics, became a dominant and limiting quantity in all of the fields of physics. A more detailed mathematical formulation then flowed from two basic postulates:

1. No experiment performed within a laboratory can determine the state of absolute uniform motion of that laboratory.
2. The velocity of light in vacuum is independent of any uniform motion of the light source or observer measuring that speed.

It is the analytical elaboration of these postulates that sets the framework for all the applications of special relativity.

Let me note at the outset that the application of special relativity most often cited does *not* in fact depend on the validity of special relativity: this is the exploitation of nuclear energy. There is no difference in principle between *chemical* reactions, such as

$$2H_2 + O_2 \rightarrow 2H_2O + \text{excess energy} \qquad (1)$$

and *nuclear* reactions, such as the fusion of light nuclei

$$^2_1 D + {}^3_1 T \rightarrow {}^4_2 He + {}^1_0 n + \text{excess energy} \qquad (2)$$

or the fission reactions of the heavy elements

$$^1_0 n + {}^{235}_{92} U \rightarrow {}^{140}_{54} Xe + {}^{94}_{38} Sr + 2{}^1_0 n + \text{excess energy} \qquad (3)$$

The chemical reaction (1) involves changes in the electron shells surrounding the nuclei, while the latter two reactions involve the nuclei themselves. Nuclear reactions release a millionfold more energy per re-acting atom than do chemical reactions. Through his celebrated equation $E = mc^2$, Einstein demonstrated that the energy release in reactions such as these corresponds to a change in mass of the reacting constituents. But this is equally true whether we are talking about ordinary chemical en-ergy or nuclear energy. However, because this effect is very small indeed in chemical reactions, the principle of mass–energy equivalence is much easier to verify experimentally in the case of nuclear processes. Work on nuclear energy for peaceful or warlike purposes could have proceeded whether or not $E = mc^2$ had ever been recognized as a profound natural law. Only the existence of a sufficient energy excess and the existence of neutron multiplication (in the case of the fission reaction) are required conditions for a practical nuclear energy source to be built.

Einstein played a major role in ushering in the nuclear age through a now famous letter that persuaded President Roosevelt to initiate atomic-bomb research with high priority. However, since the exploitation of nu-clear energy does not depend upon special relativity, I will not go into any of the enormous problems, on the one hand, and opportunities, on the other, that now face humankind in managing nuclear energy wisely.

Special relativity modifies the laws of classical mechanics, if the speed of the objects involved becomes comparable to the speed of light. This happens if the total kinetic energy of such objects approaches or exceeds their so-called rest energy. What is meant by "rest energy" is the amount of energy that would be released if the objects were completely annihi-lated in a state of rest. As an example, if a slow negative electron and its antiparticle (called a positron), each having a rest energy of about one-half million electron volts, clash and annihilate, then the total energy of the resultant fragments is 1 million electron volts (1 MeV). A million electron volts is a unit of energy; it is the amount of energy that would be gained by a particle carrying the same electric charge as an electron when it is accelerated by an applied voltage of 1 million volts. Thus, if elec-trons are used in engineering devices that have applied voltages ap-proaching one-half million volts, then special relativity must enter into the design of such devices. Similarly, the rest energy of a proton is close to 1000 million electron volts, or 1 GeV. Thus, any device that involves protons at energies approaching or exceeding 1 GeV requires special rela-tivity as an essential element in its design.

In modern engineering, there are large families of electronic devices that indeed require voltages above hundreds of kilovolts to operate in the desired manner. Thus, devices such as klystrons, high-voltage television tubes, electron microscopes, and electron accelerators for cancer therapy all require engineering analysis using special relativity in order to make them work.

A typical example is the high-powered klystron, a picture of which is shown in Figure 1. The engineer designing such tubes uses the formulas of special relativity just as he does such information as the properties of materials.

Let me give you one simple example of such design work: A klystron is basically an amplifier—it receives an electrical input signal at low voltage and low power and converts it into an output signal that is similar in shape but of much higher intensity. This is accomplished by having the weak input signal modulate the flow of a powerful beam of electrons, after which the modulated beam passes to an output structure that extracts the highly amplified information. Relativity enters into the design of this device in two fundamental ways: First, the speed of the electron stream cannot exceed the speed of light; therefore relativistic equations are essential in calculating the timing of the information imparted to the electron beam. Second, electrons in the beam repel one another, and for this reason there is a limit to the number of electrons that can be present in the stream without blowing it apart. Relativity ameliorates this effect in a manner that can be viewed consistently both by a real observer who analyzes the behavior of the tube in the laboratory and by a hypothetical observer who rides along the electron beam. The observer in the laboratory would detect not only the electric *repulsion* among the electrons, which tends to explode the stream, but also a magnetic *attraction* between the parts of the beam, which behave like wires carrying parallel electric currents. These two forces approach one another in strength as the velocity of the beam electrons approaches that of light. However—and here relativity considerations are important—one can still analyze this situation by imagining that one is riding along with the beam and that the electrons thus appear to be at rest. If the equations of special relativity are correctly applied, then, in the "frame" that is moving along with the electron beam, no magnetic effect is seen, and the electric effect is greatly weakened. This simple conclusion is consistent with the much more complicated analysis that is necessary when the problem is treated in the "normal" reference frame, in which the device itself is at rest.

There is a very fundamental reason why relativity enters more and more into engineering devices; this reason stems directly from the uncer-

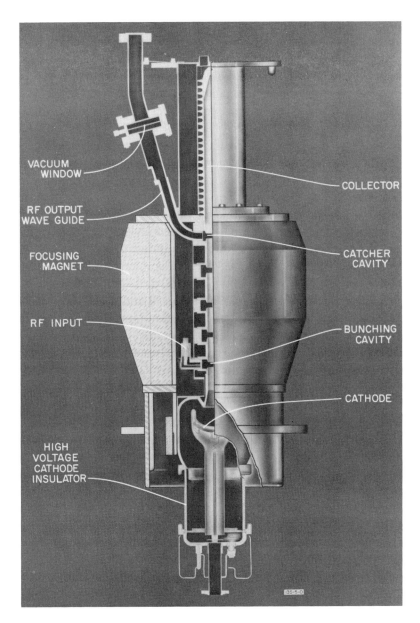

FIGURE 1. *A cutaway view of a high-power klystron. This device is about 1 meter long and operates at a frequency of about 3 gigahertz. It can generate a peak output power of about 30 megawatts and an average output power of about 30 kilowatts.*

tainty principle of quantum mechanics. This principle states that the resolution in space to which an observer can observe a given object is inversely proportional to the impulse (or what is technically known as momentum transfer) that is given to the object that is to be observed. In other words, as one wishes to discern smaller and smaller objects, the "kick" that must be given to that object must grow larger and larger. Thus, an optical microscope that involves light photons of energy of 1 electron volt or so is adequate to see objects in the submicrometer region, meaning somewhat below one-thousandth of a millimeter. In order to see biological specimens of smaller size down to macromolecules, and even individual molecules, electron microscopes having ever-increasing energies must be used. Recently, electron microscopes having accelerating voltages above 1 million volts have been successfully built and used, and they can discern objects as small as one ten-millionth of a millimeter. The ultimate microscopes are, of course, particle accelerators, which I will discuss later: these look at matter down to dimensions of 10^{-15} centimeter.

Even electron microscopes developed before World War II operated at voltages for which relativity is essential in designing the various electromagnetic lenses that constitute their focusing elements. It is interesting to note that the classical prewar engineering textbook on the design of electron microscopes and electron optics, by Zworykin and Morton, gave what at that time was probably the most extensive discussion of relativistic orbit dynamics. It is quite amusing that the formulas used in that textbook were given in the "cookbook" fashion not uncommon in engineering texts, simply telling the designer what to do without being very explicit about the physical reasons and explanations that underlie the formulas. Accordingly, the rather extensive relativistic electron-optics considerations used by electron-microscope designers went unnoticed by physicists. Therefore, much of what was written by Zworykin and Morton, and was applied by electron-microscope engineers before World War II, was rediscovered and rederived from fundamental principles by physicists designing particle accelerators after the war.

Special relativity is, of course, a totally indispensable took for the design of high-energy particle accelerators. These come in various sizes and shapes—linear and circular—and they can and do accelerate the various charged particles that are components of the atom—electrons, protons, and heavy ions. In particular for use in particle physics, energies are required that exceed by a very large factor the rest mass of the particle accelerated. For instance, the Stanford Linear Electron Accelerator accelerates electrons with a rest energy of one-half million electron volts to an energy well above 20,000 million electron volts. According to the laws of special relativity, these electrons are more than 40,000 times heavier when they emerge than they are when at rest. Their speed never becomes precisely equal to that of light, but—again according to the relativistic rules—the final speed is within one part in 3 billion of the speed of light.

Incidentally, the fact that the speed of the accelerated electron beam closely approaches that of light but never exceeds it can be verified experimentally. An experiment has been done in which a beam of light and the electron beam traveled together along the 2-mile-long pipe, and their times of departure and arrival were measured to an accuracy of approximately 10^{-12} second, which is one-millionth of a millionth of a second. To within the accuracy of the observation, they arrived at the same time.

I would like to make a few remarks about radiation of electromagnetic energy from fast charged particles. In general, when charges are accelerated—that is, if they do not move with uniform speed in a fixed direction—they radiate electromagnetic energy. A radio antenna is essentially a carrier of such charges that go back and forth, thereby producing radio waves. Therefore, one would judge that, when particles are speeded up in an accelerator, the losses that would occur due to just such radiation would have to be reckoned with. This is true, although the severity of the problem depends drastically on what particle is being accelerated and how the acceleration is done. The heavier the particle, the less the radiated power. For this reason, the matter of radiated energy does not as yet need to be taken into account for machines accelerating protons to energies thus far attainable. On the other hand, the matter is indeed serious for the much lighter electrons. Here, however, we have an interesting phenomenon: If you impart energy to electrons by accelerating them in a straight line, then, once the speed of light is approached, most of the energy becomes an increase in mass and the velocity changes only very little. Therefore, the accelerating force effectively pushes against what appears to be a rapidly increasing inertia, and there is little acceleration. Radiation from electrons accelerated in a linear accelerator is therefore negligible. This is not the case when electrons are confined by magnets to circular orbits and swing around as their energies increase. In that case, the acceleration is the ordinary centripetal acceleration that applies to any object moving in a circle, be it a person on a merry-go-round, microbe in a centrifuge, or an electron in a circular orbit. Therefore, the radiation from electrons in circular machines is a serious problem indeed, and, as

FIGURE 2. *Aerial photograph of the Stanford Linear Accelerator Center. The 2-mile-long accelerator extends from background to foreground. After acceleration, the electron beam can be directed to any of several different experimental areas located in the larger buildings in the foreground.*

the energy increases, more and more radio-frequency power is required to compensate for this radiation loss. We will discuss this problem in somewhat greater detail a little later. This is another example of relativity consideration playing a controlling role in the design of accelerators or even in the basic choice as to which accelerator serves the need in question.

Let me turn now specifically to the linear accelerator at Stanford University, which is shown in Figure 2. The particles are injected at low energy and are then carried along in synchronism with a traveling electric field, gaining energy as they go. The inner diameter of the accelerating pipe is about 2 centimeters, while the length, as already mentioned, is about 2 miles, or 3 kilometers.

The question comes to mind as to how it is possible to keep the electron beam within this narrow pipe for such a long distance. Again, we can provide the answer in either of two ways, using relativistic arguments.

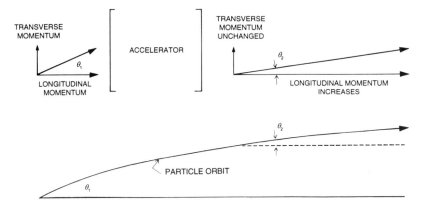

FIGURE 3. *During the acceleration process, the longitudinal momentum of the beam electrons increases continuously, while the transverse momentum remains the same. The net effect is to decrease the angle made by the beam particles with respect to the accelerator axis from an initial value of θ₁ to a final, much smaller value of θ₂.*

The first is as follows: As the particle accelerates, its longitudinal momentum is being increased continuously. You may recall that momentum is the product of mass and velocity. In this case, the velocity is almost constant and nearly equal to the speed of light, while the mass continues to increase in accordance with the energy–mass equivalence principle of Einstein. It turns out that, on the average, there are no radial forces acting on the electrons in the pipe; therefore, the transverse momentum does not increase but stays constant throughout the acceleration process. This is shown in Figure 3. As shown, the angle that the beam makes with respect to the axis continually decreases. Thus, the beam becomes increasingly collimated or confined, and for this reason it does not spread out substantially during acceleration.

There is another—and more striking—way to visualize this situation by using the arguments of special relativity. Suppose we could observe this process while traveling along with the accelerating electrons. From this vantage point, we would observe that the length of each of the individual segments of the accelerator became progressively shorter as our speed, relative to the pipe, approaches the speed of light. In fact, the 2-mile-long accelerator would appear to have a length of only about 30 inches as we rode along with the electrons! This means that the task of properly injecting electrons into the 2-mile Stanford machine is simply to

FIGURE 4. *The SPEAR storage ring at SLAC is shown at the right in this aerial photograph. In this device, counterrotating beams of electrons and positrons of energies up to 4 GeV per beam collide at two different locations around the ring circumference.*

aim them accurately enough so that they will hit a target only 2 centimeters in diameter at a distance of only 30 inches, and any fool can do that!

Therefore, whether you are looking at this problem from the point of view of an observer at rest relative to the accelerator structure or from the point of view of an observer traveling along with the electrons, the ease of confining such an electron beam can be understood in a consistent manner. Thus we see that, to qualify as a designer for an electron linear accelerator, you must understand something about special relativity, and this is true also for the designers of circular accelerators, which are used, among other things, for accelerating protons to the highest energies yet achieved by artificial means.

A recent development in the accelerator field has been the confinement of high-energy electrons in circular storage rings after they have been previously accelerated by an accelerator. Figure 4 shows one such installation. These storage rings have become powerful tools for high-energy elementary-particle physics, and they have proved to be very useful in other applications. In particular, the circulating beam of electrons has the characteristic that it emits high-energy x rays of unprecedented intensity. We have discussed this process previously as a serious limitation faced by the designer of circular machines for accelerating or storing electrons. However, one man's pain can be another's pleasure: This radiation can be used for whatever useful purposes people may have for x rays. This type of radiation goes under the name of *synchrotron radiation*. It might be interesting to consider, again using the arguments of relativity, why such high-energy x rays are emitted.

This radiation process is shown schematically in Figure 5. You will remember that traveling around in a circle is equivalent to being accelerated continuously toward the center of the circle. In other words, if you are not accelerated at all, you travel in a straight line, but if you are continuously accelerated from a straight line toward a fixed point, then the result is that you travel in a circle. It is also characteristic of the laws of electricity that, if a charge, such as an electron, is accelerated, it radiates electromagnetic radiation. For instance, a radio antenna is simply a piece of wire in which electrons are alternately accelerated and decelerated, and this motion of electric charge produces radio waves, which can be received at long distances. Thus, an electron traveling in a circle should behave like a little antenna that emits radio waves in a pattern that is exactly the same as would be produced by a short-wire antenna pointed toward the center of the circle. But does it?

In practical electron-storage rings, the frequency with which the electron goes around is roughly 1 million times per second, and from this one might surmise that the frequency of the radiation produced would also be 1 million cycles per second, or 1 megahertz, which is in the middle of the AM radio broadcast band of the electromagnetic spectrum. However, you may also remember that radio waves have a very much longer wavelength and a lower frequency than x rays. Why, then, does an electron traveling around in a circle at nearly the speed of light emit x rays rather than radio waves?

The reason can be understood as follows: First, it is indeed true that, *if* you were traveling along with the electrons, you would see a radiation pattern that corresponds to that of an ordinary wire antenna. This is shown in Figure 5. However, from the point of view of an observer who is not traveling with the electron but is at rest relative to the storage ring,

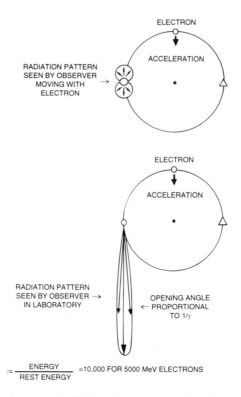

ELECTRON

ACCELERATION

RADIATION PATTERN
SEEN BY OBSERVER
MOVING WITH
ELECTRON

ELECTRON

ACCELERATION

RADIATION PATTERN
SEEN BY OBSERVER →
IN LABORATORY

OPENING ANGLE
← PROPORTIONAL
TO 1/γ

$$\gamma = \frac{ENERGY}{REST\ ENERGY} = 10,000\ FOR\ 5000\ MeV\ ELECTRONS$$

FIGURE 5. *The radiation pattern emit-ted by an electron in a storage ring, as seen by an observer moving with the electron (above) and by an observer in the laboratory (below). In the latter case, the opening angle of the emitted radiation is proportional to 1/γ, where γ is the ratio of the electron's total energy to its rest energy. For electrons with en-ergies of 5000 MeV, γ ≅ 10,000.*

the radiation pattern is thrown forward into a very narrow cone. The an-gle of this cone is just the inverse of the ratio of the total energy of the electron to its rest energy. This ratio is given by the symbol γ (gamma) in relativity. Thus, the wave striking an observer is not continuous but is a short pulse emitted for only that fraction of the period of revolution of the

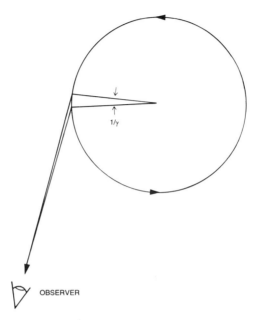

FIGURE 6. *As the radiation cone sweeps past the observer, the emitting elec-*
tron moves through the small angle $1/\gamma$. If the electron were traveling at ex-
actly the speed of light, then the radiation emitted at the beginning and at
the end of the short time interval would arrive at the observer at exactly the
same time. However, since the electron does not quite travel at the speed of
light, there is a small difference, and special relativity can be used to show
that the time of arrival of the two radiation pulses at the observer is short-
ened relative to the time interval of emission by the ratio $1/\gamma^2$. This means
that the time interval during which the observer receives radiation is
shorter by a factor of $1/\gamma^3$ than the time it takes for the electron to go
once around its circular orbit. Consequently, the wavelength of the radia-
tion is reduced by the same factor, and its frequency is correspondingly in-
creased by the factor γ^3.

electrons that corresponds to the ratio of these two energies—that is, the
fraction $1/\gamma$.

But this is not all. We also must consider the fact that, during the time
when the observer receives this short pulse of radiation, the electron itself
has moved. As is shown in Figure 6, the electron has moved through this
small angle, $1/\gamma$, as the radiation cone sweeps past the observer. If we
consider light pulses that are emitted at the beginning and at the end of

that time interval, we find that these two light pulses will arrive at the observer at almost the same time. In fact, if the electrons traveled exactly at the speed of light, the pulses would arrive exactly at the same time because of the extra distance that the first light pulse had to travel to reach the observer. However, since the electron does not quite travel at the speed of light, there is small difference, and special relativity can be used to show that the time of arrival of the two light pulses at the observer is shortened relative to the time interval of emission by the ratio $1/\gamma^2$. This means that the time during which the observer receives light is shorter, by a factor of $1/\gamma^3$, than the time it takes for the electron to go once around its circular orbit. In consequence, the wavelength of the radiation is reduced by the same factor, and its frequency is correspondingly increased by the factor γ^3.

For practical electron storage rings, the value of γ can be very high. As an example, if we have electrons with an energy of 5000 MeV going around in a circle, then the mass or energy of such particles is 10,000 times the rest energy: $\gamma = 10,000$, and $\gamma^3 = 10^{12}$, or a million million. This huge factor shifts the emitted frequency from the radio broadcast band into the x-ray region. Electron storage rings are thus very practical devices for producing extremely intense, high-energy x rays, which are being used in a large variety of applications: research in biology and medicine, production of integrated circuits and other semiconductor devices, determination of the structures of complex atoms and molecules, and so on. So the design of x-ray sources based on synchrotron radiation again clearly requires the engineer to be fluent in the principles of special relativity.

A great deal of folklore has developed around what is commonly called "the twin paradox" of special relativity. If two clocks keep identical time when they are at rest relative to one another, then the predicted result of relativity is that, if one of them is set in motion and then returns, it will show less elapsed time than the one that remained at rest. In human terms, if two identical twins were truly identical, and if one of them went on a trip and came back, then he would be younger than his stay-at-home brother. This prediction raises two questions: First, is it true? And, second, is it really a paradox—and, if so, what is the paradox? We can answer these questions unequivocally: *Yes*, it is true; and *no*, it is not a paradox, because we understand it. Indeed, the twin taking the two-way trip would age less rapidly, as measured on the biological time scale of his brother, and the amount of this difference depends on the ratio of his travel speed to the speed of light. The effect is very small indeed, when

one of them went on a trip and came back, then he would be younger than his stay-at-home brother. This prediction raises two questions: First, is it true? And, second, is it really a paradox—and, if so, what is the paradox? We can answer these questions unequivocally: *Yes*, it is true; and *no*, it is not a paradox, because we understand it. Indeed, the twin taking the two-way trip would age less rapidly, as measured on the biological time scale of his brother, and the amount of this difference depends on the ratio of his travel speed to the speed of light. The effect is very small indeed, when one considers ordinary speeds. For example, if the twin's trip proceeds at 600 miles per hour, then the difference in aging rate between the twins will be about 1 part in a million million, or two-thousandths a second in a human lifetime, hardly an observable amount. But what is most puzzling is the principle of the thing: If the twin comes back and if the effect *were* observable, then the twins would no longer be identical, because one of them would have aged less than the other. Therefore, the final outcome is, in principle, different if A has gone on a trip and B has stayed at home than it would have been had B gone on a trip and A stayed at home. This seems to violate what one naively believes to be the essence of special relativity—namely, that you cannot determine absolute motion, since there is no fixed reference frame that defines a state of rest. This conclusion, however, does not apply to this particular instance. Special relativity indicates only that reference frames moving at a *constant* relative velocity cannot be distinguished. It does *not* say that you cannot distinguish between accelerated and nonaccelerated states of motion. For one twin to take a trip and to return means that he has to be accelerated during part of his travel because, during his turn around, his speed obviously must be reversed. Therefore, the one who has aged less is the one who has been in accelerated motion; thus, the twin paradox is not actually a paradox but constitutes only one of the many experiments that can tell the difference between what professionally is called an inertial frame and a noninertial frame—that is, a frame that is in uniform accelerated motion relative to that "inertial" frame defined by the preponderance of the masses in the universe.

In order to describe what is going on here, either from the point of view of one twin or from that of the other, we have to go beyond the boundary of special relativity and turn to general relativity, which offers a consistent method for dealing with accelerated states of motion also. However, special relativity does give the "right answer," if a phenomenon is *described* in the inertial frame, and, in such a frame, the "stay-at-home" twin has aged more than his itinerant brother.

Now this may all be interesting enough, but why do I mention it in a discussion of relativity and engineering? The answer is that, although the twin paradox is not observable with actual human twins, there is a very strong and demonstrable effect when one talks about phenomena on an atomic scale. We do indeed have clocks that can be put into rapid motion, be sent around in circles, and come back home. Such clocks are radioactive atoms or particles that can decay radioactively into lighter fragments. Such unstable particles have built into them a basic clock that determines the probability that the particle will disintegrate. For example, the mu meson, one of the particles in modern physics that occurs copiously in secondary cosmic rays reaching the earth, has a "mean life" of 2×10^{-6} second. That is, on the average, it will decay after two-millionths of a second. This lifetime has been measured very precisely. If one takes a beam of such particles and captures it in a ring-shaped electromagnet, the particles will go around and around until they decay, and we can thus keep track of their numbers over time.

Such mu-meson beams can be produced with a velocity approaching the speed of light. For instance, if one stores mu mesons with an energy of 3 GeV (3 billion electron volts) in a ring magnet, as has been done during the past few years in experiments at CERN in Geneva, the so-called time dilation factor that is observed is about 30. Thus, a mu meson at rest will have an average lifetime 30 times shorter than those traversing circular orbits at 3 GeV. This effect agrees very precisely with the prediction of special relativity theory. The twin paradox is thus both real and precisely measurable in practical situations.

These "practical situations," however, have thus far been confined to radioactive clocks. There has been much speculation in science fiction about space travel, in which passengers go out into space and then return, having outlived their contemporaries. Can this be made to work? In practical terms, the answer is *no*. Although space vehicles are useful platforms to mount highly sensitive experiments to examine various features of the general theory of relativity, space travel by human beings is one field for which the engineer does not have to worry about special relativity.

Particle Substructure:
A Common Theme of Discovery
in This Century

It is not once nor twice but times without number that the same ideas make their appearance in the world.

Aristotle, *On the Heavens*

After exposure to public accounts of scientific work, the layman might conclude that new results or "breakthroughs" in science tend to supersede results that were previously known and accepted. In physics, however, this is rarely the case. Rather, the more common situation is that the range of validity of an older concept is found to be less than universal. For instance, Newtonian mechanics remains as valid for velocities that are small compared with the velocity of light as it did before the advent of relativity.

What is even more remarkable is that, despite the explosion of knowledge about nature on a smaller and smaller scale, some of the basic rules that describe the behavior of matter remain equally valid today. It is to this latter phenomenon that this talk is addressed. Specifically, I will give some examples from modern developments in particle physics that demonstrate that the fundamental rules of quantum mechanics, applied to all

---

A lecture given at Cherwell-Simon Memorial, Oxford, May 8, 1981.

forces in nature as they became understood, have retained their validity. The well-established laws of electricity and magnetism, reformulated in terms of quantum mechanics, have exhibited a truly remarkable numerical agreement between theory and experiment over an enormous range of observation.

As experimental techniques have grown from the top of a laboratory bench to the large accelerators of today, the basic components of experimentation have changed vastly in scale but only little in basic function. More important, the motivation of those engaged in this type of experimentation has hardly changed at all.

The Nuclear Atom

Let me begin this story with the classical experiments of Lord Rutherford around 1911, which demonstrated that the atom is not a continuous blob of matter but is rather a structure whose mass is concentrated in a small, central nucleus. Let me show the basic arrangement in Figure 1.

Rutherford's experiment made use of a beam of alpha particles formed simply by collimating the particle flux emitted in all directions by a natural radioactive source. This *beam* of alpha particles was then directed at a *target,* which, in the case of Rutherford and collaborators, was a thin foil of the material under study. Alpha particles were then scattered into a *detector,* which, in this case, was a florescent zinc sulfide screen. This was, in turn, viewed by a graduate student through a small microscope. Thus we see here a typical scattering experiment—beam (collimated alpha particles), target (scattering foil), and detector (florescent screen, microscope, and graduate student). You will see that all of the more recent scattering experiments to which I will refer still have precisely these same basic components.

Scattering experiments are the most common method for the physicist to analyze the constituents of particles that are not (with the techniques available at a given time) accessible to *individual* observation. A scattering experiment simply measures the probability that an incident particle is scattered at a given angle away from its initial direction and, in so being, has either suffered or not suffered an energy loss or a change in some other characteristic during the scattering process. If the energy of the incident particle after deflection is diminished only by the recoil energy of

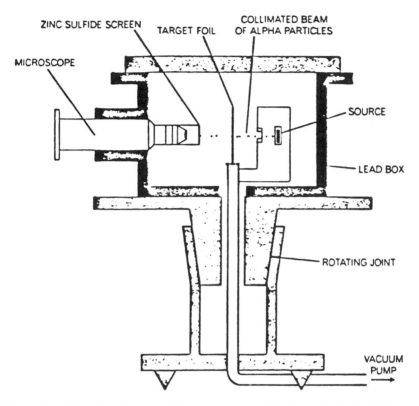

FIGURE 1. *Schematic diagram of the apparatus used by Ernest Rutherford that demonstrated the existence of the nucleus within the atom. The apparatus consists of a source of alpha particles, followed by a collimator, which forms a beam of these particles striking a target foil. After interacting with the foil, the particles then strike a zinc sulfide flourescent screen. The location of the particles on the screen is determined by a human observer viewing the screen through a microscope.*

the target particle without any other changes, then the scattering is called elastic; otherwise, it is called inelastic.

Scattering of one particle on another gives a "kick," or what the physicist calls "momentum transfer," to both particles. It is the magnitude of this momentum transfer that measures the scale to which the scattering process can give information of the structure of the particles. The relationship between the resolution Δx to which the spatial structure of the particles can be revealed by the scattering process and the momentum transfer Δp is given by Heisenberg's uncertainty principle, $\Delta p \, \Delta x \cong h/2\pi$

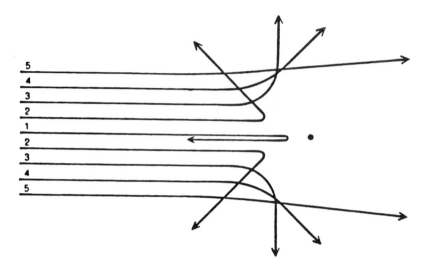

FIGURE 2. *Schematic representation of orbits of alpha particles colliding with a concentrated, charged object. Note that close collisions result in large angles of deflection, while collisions at a distance produce only a small change in direction of the incident particles.*

where h is Planck's constant. This relation, in turn, sets the practical energy required for the incident beam, if structure down to a specified dimension is to be resolved.

It is well known that Rutherford's scattering experiment led to the discovery of the nucleus—that is, the central core of the atom—which is about 10,000 times smaller in diameter than the atom itself, yet contains almost all of its mass. Figure 2 shows schematically how Rutherford reached his conclusion. It was observed that the number of particles scattered with large momentum transfer (that is, at a large angle) exceeded the amount predicted, assuming that the atom was a continuous blob of matter.

It is interesting to note that the analytical tools available to Rutherford to predict how much scattering would occur, and at what angle, had to be based on the then accepted laws of classical mechanics and classical electricity and magnetism. In other words, even though Rutherford's experiments probed matter at a scale smaller than had theretofore been accessible to experimental observation, he had to assume that the physical laws derived from large-scale observation were still valid at small distances. Thus, using the same rules to analyze atomic collisions as would be used to analyze collisions between charged ping-pong balls, the conclusions about the existence of the atomic nucleus were drawn.

In retrospect, Rutherford's conclusions were right, but his methods were wrong. We know that, at the magnitude of momentum transfer that was produced in the collisions observed by Rutherford, the laws of classical mechanics would no longer be valid, but that the laws of quantum mechanics, which were not established until more than a decade later, were to be applied instead.

However, it turns out, fortuitously, that the classical and quantum-mechanical calculations for the probability of scattering give essentially the same answer, provided the scattering is controlled by the laws of electromagnetism, and also provided that the scattered particles move at speeds well below the velocity of light. Had the forces active in Rutherford's experiment been other than those of electromagnetism, then indeed the quantitative result of Rutherford's calculations would have been incorrect.

The Scattering of X Rays

Now let me turn to another class of scattering experiment: this is the scattering of light and x rays from the atom. The scattering of light or x rays from the atom is dominated by the extranuclear electrons, not by the nucleus. The reason is that the interaction of light or x rays, which are electromagnetic radiations, is with an electric *current*; in the scattering process, such a current is produced by the recoiling particle from which the scattering takes place. If the mass of the recoiling particle is small, then a larger recoil current is produced. Thus the light electrons surrounding the heavy nucleus are the principal contributors to the scattering of electromagnetic radiation by the atom.

This type of scattering was observed first by A. H. Compton in experiments beginning in 1922. In these experiments, it was shown clearly that the dynamic properties of the incident x ray beam were described by its photon or quantum characteristics— that is, the energy and momentum of electromagnetic radiation behaved like those of a particle.

It is characteristic of the process of scattering of x rays by a free particle that the frequency and energy of the scattered quantum is shifted relative to those of the incident radiation. Particularly interesting is the fact that the shift in wavelength of the x rays in the scattering process depends

FIGURE 3. *Photograph of Jesse W. M. DuMond and his collaborator, Professor Harry A. Kirkpatrick, with their multicrystal spectrometer. This instrument was used to analyze the shape of x-ray spectra that resulted from the scattering of x-ray photons from various materials.*

only on the mass of the particle struck and on the angle of deflection of the incident radiation; it is independent of the energy of the incident x rays. The most important result from the observation of Compton scattering was the establishment of the quantum properties of the photon itself, rather than any new insight into the structure of the atom, In fact, in the original Compton scattering experiments, it would have made little difference if the electrons on which the scattering took place were free or were bound to the nucleus.

The power of Compton scattering in analyzing not only the structure of atoms but also their internal dynamics became evident with the experiments of J. W. M. DuMond and collaborators, starting in 1926, which observed the frequency of scattered photons with much higher precision. These experiments revealed not only that scattering of incident photons on atomic electrons indeed took place but also that these electrons were themselves in a state of motion. Figure 3 shows DuMond with

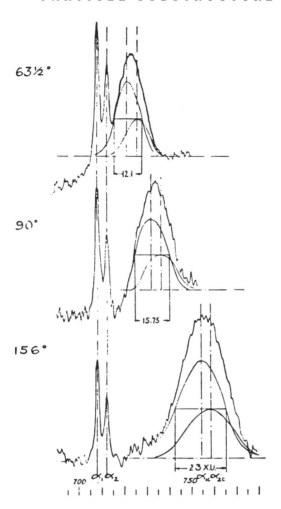

FIGURE 4. *Microphotometer traces of photographic spectra obtained in the multi-crystal spectrometer of DuMond and Kirkpatrick. The spectra show the unmodified molybdenum $K_{\alpha 1}$ and $K_{\alpha 2}$ lines, together with the Compton-shifted line, as scattered from graphite at various angles. Note that the broadening, as well as shift itself, increases with scattering angle as predicted by theory.*

his apparatus. This instrument, consisting of a complex assembly of crystals, recorded data on photographic plates. In turn, the exposure of these plates was measured by a microphotometer; Figures 4 and 5 show samples of the resulting spectra. Classically, the experiments of DuMond and collaborators could be interpreted to yield the velocity distribution of the

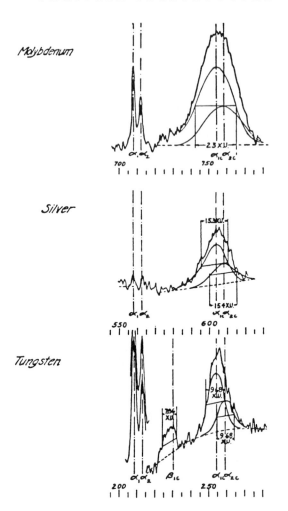

FIGURE 5. *Microphotometer traces of photographic spectra obtained in the multi-crystal spectrometer of DuMond and Kirkpatrick. The spectra show both the unmodified and Compton-shifted $K_{\alpha1}$ and $K_{\alpha2}$ lines of molybdenum, silver, and tungsten, respectively, scattered at a fixed angle from graphite. Note that the shorter the wavelength, the narrower the scattered line. Note also that the central Compton shift at fixed angle is independent of the incident wavelength.*

electrons in atoms. Figures 4 and 5 show that the dynamic width of the "scattered" line increases with scattering angle and incident wavelength, respectively, as can easily be computed from the nature of the process.

DuMond initially analyzed his scattering experiments by assuming that the electrons moved in classical Bohr orbits around the nucleus. In that case, Compton scattering is simply modified by the Doppler shifts produced by the motion of the struck particle. Quantum mechanically, one must analyze modified Compton scattering by describing the momentum distribution of the extranuclear electrons on a probabilistic basis, without talking about actual orbits. Both types of analysis give similar results. In a certain sense, these experiments are complementary, in that Rutherford scattering is elastic scattering on the nucleus, while Compton/DuMond scattering is *inelastic* scattering on the atom as a whole but *elastic* scattering on the atomic electrons. The former constitutes the basic discovery of the structure of the atom, while the latter constitutes measurements of a dominant feature of the dynamics of that structure.

The preceding discussion has dealt with scattering by a *single* electron. Naturally, if more than one electron is involved, the situation becomes more complex. This complication arises if the wavelength of the incident radiation is comparable to the size of the distribution of the electrons. In that case, the scattering of the light by the individual electrons produces interference effects similar to those observed when visible light scatters off the individual elements of a diffraction grating. In general, one can separate the observed scattering into two factors: one is the term that governs the probability of scattering of the x rays from an individual electron, and the other is the factor that measures the interference effect due to the multiplicity of scattering sources; the latter is known as a "form factor." We will meet this type of factorization again when we talk about scattering at much higher energies.

High-Energy Electron Scattering

Now let us switch from x rays to electrons for the incident beam, go forward by about four decades, and increase the energy of the particles to be scattered by a factor of about 1 million. Scattering again can be both elastic and inelastic. Elastic scattering yields information on the radius and general distribution of charge within the proton and neutron. It was experiments with elastic electron scattering carried out during the 1950s, for which Robert Hofstadter received the Nobel Prize in physics in 1961, that determined these basic parameters. The fact that the proton has a finite radius indicates in itself that the proton cannot be an ultimate constituent of

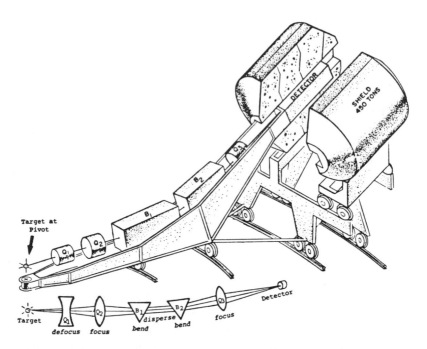

FIGURE 6. *Diagram of the magnetic spectrometer at SLAC capable of analyzing particles with momentums of up to 8 GeV divided by the velocity of light. The incident beam strikes a target around which the spectrometer rotates. The spectrometer consists of focusing elements labeled "Q" (for quadruple lens) and deflecting magnets labeled "B" (for bending magnets). They analyze the particles scattered in the target, which are then identified and registered in the detector.*

matter but rather must have a substructure of some kind. The general nature of this substructure was revealed through experiments at SLAC, beginning in 1967, that concentrated primarily on inelastic rather than elastic scattering—that is, interactions in which the proton is disintegrated in consequence of the scattering process.

Figure 6 shows schematically, and Figure 7 shows photographically, the apparatus that was used at SLAC to study both inelastic and elastic scattering of electrons of energy up to 20 giga-electron-volts (GeV) on hydrogen targets. Note that the basic components of this apparatus are the same as the ones used by Rutherford. We have an incident *beam* of charged particles; we have a scattering *target,* here consisting of a chamber containing liquid hydrogen; and we have a *detector* composed of

FIGURE 7. *Photograph of the spectrometer shown diagrammatically in Figure 6, together with a second instrument capable of analyzing particles with energies of up to 20 GeV divided by the velocity of light.*

magnetic spectrometers that measure precisely the angle of scattering and the energy and nature of the scattered particle. Thus, the basic components of a scattering experiment have remained the same throughout this century, as has the spirit of the investigation: Rutherford wished to investigate the substructure of the atom, while the SLAC experiments investigated the substructure of the proton and neutron, which had been established as the fundamental building blocks of the nucleus discovered by Rutherford. While the basic nature and motivation of the experiments have not changed, the scale indeed has. This is a consequence of the uncertainty principle: In order to study matter at smaller dimensions, the transfer of momentum must be proportionately large. Roughly speaking, the proton has a diameter 100,000 times smaller than the atom, and therefore the energies must be increased roughly in that proportion.

In analyzing inelastic scattering, the experimental physicist has several tools at his command. He can examine the fragments that are ejected as a result of the disintegration process, or he can examine the loss of energy of the incident electron during the scattering process; it is this loss in energy that presumably corresponds to the energy of creating and propagat-

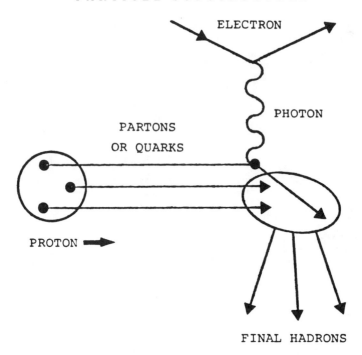

FIGURE 8. *Diagram of the deep inelastic electron-scattering process on the proton. It is assumed that protons are composed of three partons, or quarks. The electron "virtually" emits a photon during the scattering process that interacts with one of the three partons. The three partons interact after the scattering process and recombine, forming a combination of final hadrons.*

ing the "ejecta" from the proton. Most of the relevant information regarding the electron-scattering process was obtained by experiments that looked only at the scattered electron.

If we assume that the proton has constituents that are pointlike, or at least have a radius very much smaller than that of the proton itself, then an inelastic scattering process can be envisaged as the totality of elastic electron scattering from each of these pointlike constituents. This is shown in Figure 8. One can easily derive geometrical relations between the direction and energy of the incident and scattered electron under a number of plausible assumptions. Specifically, by arguments exactly analagous to those applied to x-ray scattering by the atom, one can show by simple mathematics that the energy and angle distribution of the outgoing electron should be a product of two factors. The first factor is the distribution corresponding to scattering from a single, pointlike object. This first factor can be calculated with confidence from theory, since the forces governing the interaction between an electron and a pointlike charged object are almost purely electromagnetic. In essence, electron scattering is thus "exploring unknown structures with known forces." The second factor, again called the "form factor," is a function characteristic of the distribution of the pointlike objects within the proton. As long as the basic model of the process is correct (that is, scattering occurs from pointlike objects coupled by forces that can absorb transverse momenta up to a specified limit), then the form factors should depend only on a certain dimensionless ratio defined by the kinematics of collision. This ratio can be identified with the fraction of the momentum within the proton that is carried by the struck pointlike object. The assumption made here is that binding among these objects is negligible in a consideration of their motion.

This simple dependence of the "form factors" is known as "scaling," and it can be considered to be an indication of the fact that such pointlike constituents might indeed exist within the proton. Figures 9 and 10 show how well this simple description agrees with the experimental data. It can be seen that the agreement is good but not perfect. Naturally, the total range of variables over which the correctness of this model can be tested is limited both by the energy of the available beams and by the data rates that can be recorded.

Another, more dramatic demonstration of the existence of pointlike particles can be made by comparing elastic and inelastic scattering. This is shown in Figure 11. It is seen that the probability of elastic scattering on the proton as a whole falls off much more rapidly with increasing momentum transfer than does the probability of scattering that affects presumably only single, pointlike objects. You will recall that this is precisely what Rutherford observed in regard to the nucleus within the atom. By this analogy, we see at least a strong indication that the proton and neutron do indeed have pointlike constituents, which were first dubbed "partons" by Richard Feynman. These are now recognized to be identical to the pointlike "quarks" whose existence had been postulated through a completely different line of reasoning; this was to explain the great number of different particle states that had been discovered by the high-energy physicist, the systematic relations among their masses, and the rules that govern their conversions into one another.

More careful examination of the inelastic-scattering process has shown

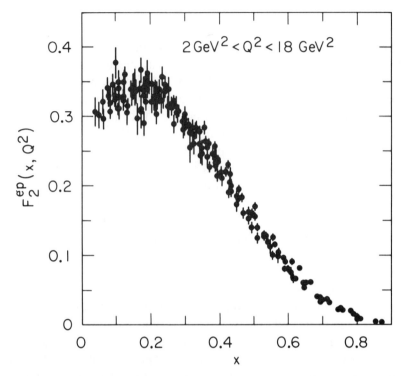

FIGURE 9. *The experimental evidence for "scaling" in deep inelastic electron-scattering on the proton. The graph shows the form factor plotted as a function of the parameter* $x = q^2/2M\nu$, *where q is the momentum transfer, M is the mass of the proton, and ν is the energy difference between the incident and the scattered electron. The x on the horizontal axis is the fraction of the momentum carried by the "parton" struck by the incident electron, as measured in a frame in which the proton is in rapid motion.*

that scaling is not exact. This conclusion can be drawn both from the highly precise experiments at SLAC, using electron energies up to 20 GeV or so, and from the less precise experiments at Fermilab near Chicago, and at CERN in Geneva, using beams of higher-energy muons.

Experiments that measure the deviation from scaling constitute a valuable tool for examining the nature of the forces among the partons or quarks. According to modern concepts, these quarks interact through the exchange of certain objects called "gluons." The interaction between these particles and the quarks is measured through a coupling constant, whose strength, in turn, determines the deviation of the data from the ideal scaling relationship, which is derived assuming the quarks to be

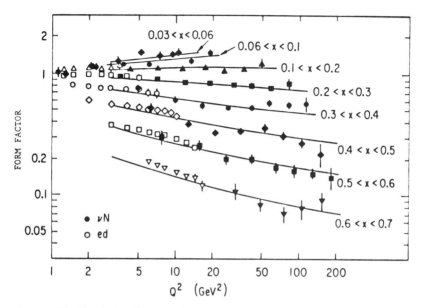

FIGURE 10. *Deviation from scaling. This graph shows that there is a weak dependence of* q *of the form factor in deep inelastic scattering, where the character of that weak dependence is different for different values of* x. *Interpretation of this type of information gives valuable information on the interaction between quarks within the proton.*

"free." Thus, the deviation from scaling measures the strength and character of this basic force.

Note here the analogy with the x-ray experiment that I described earlier. The first experiments of Compton demonstrated that x-ray photons indeed bounced off the electron component of the atom, which behaved nearly as if they were "free." The more refined experiments of DuMond and collaborators showed *deviations* from pure "free" Compton scattering and thus gave evidence of the dynamics of the atom—in particular, the momentum distribution of electrons, in turn derived from the strength of the interaction between the electron and the nucleus. In the recent experiments, the existence of scaling as such indicates the existence of partons or quarks, but it is the *deviation* from ideal scaling that gives valuable evidence of the dynamics that binds the quarks together to form a proton or neutron. In the case of the atom, the controlling basic forces are purely electromagnetic, and they can be described by what is now known as quantum electrodynamics. This theory is well understood, and its validity

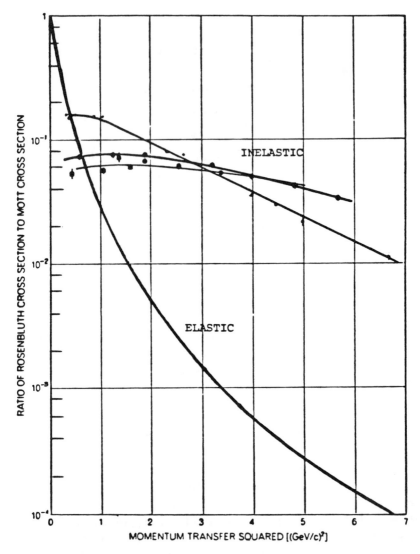

FIGURE 11. *Comparison of inelastic and elastic electron–proton scatter-ing. This graph shows the ratio of the observed scattering to the point par-ticle cross section for elastic and inelastic events plotted as a function of the square of the momentum transfer. Note that inelastic scattering falls off much more slowly with momentum transfer than does elastic scatter-ing. This indicates that inelastic scattering appears to take place on pointlike objects within the proton.*

has been demonstrated experimentally over distances ranging from many earth diameters down to about 10^{-16} centimeter. The analagous theory of the strong forces acting between quarks and carried by gluons is known as quantum chromodynamics: it is now being formulated, although many outstanding questions remain to be answered.

However, irrespective of the nature of these forces, the basic "rules of the game"—namely, quantum mechanics and relativity—have continued to apply throughout. I will now turn to a similar story in regard to the study of the simplest possible bound-state systems, which consist of just two particles: a particle and its antiparticle.

Particle–Antiparticle Bound States

One of the predictions of relativistic quantum mechanics is that, for each of the particles found in nature, there should also be a corresponding antiparticle—that is, a particle with not only its electric charge reversed but also certain of its other characteristics reversed as well. This, in turn, led to the expectation that a charged particle and its antiparticle could combine to form a quasi-stable system, essentially a planetary system in miniature. Specifically, it was predicted, soon after the discovery of the antielectron, or positron, that there should be an entity now known as "positronium," which is the bound system of an electron and a positron. Similarly (again looking ahead many decades), if indeed quarks are fundamental constituents of nuclear matter—or, more accurately, of all hadronic matter—then there should also be "quarkonium" systems of various kinds, each consisting of a quark and an antiquark. Such objects are the simplest bound systems one can imagine. One would therefore presume that the positronium system constitutes an ideal test object to examine the validity of the theory of electromagnetic forces, while the quarkonium systems might be a similar laboratory for examination of the forces that govern the behavior of the constituents of nuclei or other strongly interacting particles.

The first experiments of positronium were done in 1951 by Martin Deutsch of MIT. A very large volume of work on positronium has been done between that time and today. This work has resulted in a complete and accurate tabulation of many energy levels of positronium and extremely precise measurements of the transitions between them. Figure 12 shows a diagram of these energy levels and of the numbers that go with

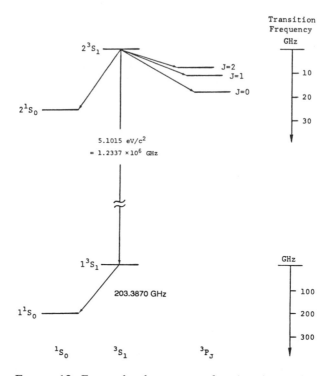

FIGURE 12. *Energy-level spectrum of positronium, using the ordinary spectroscopic term designation giving spin multiplicity, orbital angular momentum, and spin angular momentum, respectively. The spectrum is plotted in three different scales: (1) one scale, shown in the upper right-hand corner, corresponding to the transitions among levels of principal quantum number two; (2) a second scale, shown in the lower right-hand corner, corresponding to the transition between levels of principal quantum number one; and (3) a third scale, on which the transition between the energy levels of principal quantum number 1 and 2 is shown, that is larger by 5 orders of magnitude than the transitions within each principal quantum number.*

them. These numbers constitute one of the most fundamental tests of the validity of quantum electrodynamics. For instance, the transition energy between the two lowest levels of positronium in theory is given by 203.40 gigahertz, with an uncertainty of $\pm$ 0.01, while the experimental measurements give 203.3870.

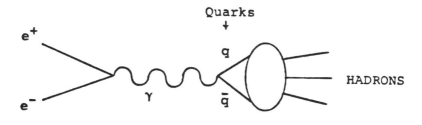

FIGURE 13. *Diagrammatic representation of electron–positron annihilation, resulting in final hadrons in the upper diagram and final lepton pairs in the lower diagram. Since hadrons are composed of mixtures of quarks, the primary process in the upper diagram is creation of quark pairs via electromagnetic interaction carried by the intermediate photon* γ. *Hadrons are then produced subsequently through forces between the quarks. In the lower diagram, in contrast, it is shown that leptons are produced through a purely electromagnetic process.*

This type of spectacular agreement is characteristic of the many experiments that have demonstrated that, as long as only electromagnetic forces are affecting the process, quantum electrodynamics fully explains quantitatively all the observed phenomena.

The most sensitive tests of the validity of quantum electrodynamics at the smallest distances of interaction come from experiments at the highest energies. If electrons and their antiparticles are not bound together as they are in positronium but rather collide at the highest energies available in the laboratory, then a number of things can happen, as indicated in Figure 13. The best way to produce such collisions is in high-energy electron–positron storage rings. Figure 14 shows an example of such an installation at SLAC.

FIGURE 14. *Aerial photograph of the housing containing the SLAC electron–positron colliding-beam storage ring SPEAR. The two buildings inserted into the ring contain the experimental apparatus detecting the results of electron–positron annihilation.*

Digression: The Growth of Accelerators

Let me digress here to give some indication of how the enormous gap between the early, low-energy experiments of Rutherford, Compton, DuMond, and Deutsch and the recent, high-energy experiments on electron scattering and electron–positron collisions has been bridged. In this talk, I am emphasizing the similarity in concept, but the dissimilarity in scale, between the early and recent experiments. The actual *pace* of this progression has been defined by the evolution of the technology of high-energy accelerators and colliders. Figure 15 is an update of a chart, originally devised by S. Livingston, which shows how the energy available through the use of accelerators has evolved over time. The pattern is indeed dramatic: the energy of accelerators has increased by a factor of 10 approximately every 7 years ever since the 1930s. This has been achieved not simply by building larger and larger accelerators of a single type but,

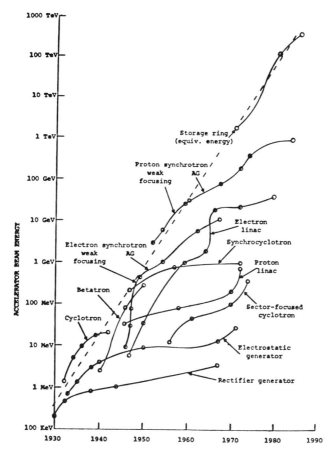

FIGURE 15. *Growth over time of the energy attainable through particle accelerators and storage rings. Note that this growth is achieved through a succession of machines employing different technologies.*

rather, by a succession of new technologies that were invented whenever an old technology began to flag in its ability to reach higher energies. Thus the growth in size and cost of accelerators has not been anywhere near as large as the increase in scientific potential. I cannot describe here in any great detail the succession of inventions and technological advances that have made this evolution possible. Let me name just a few.

The early accelerators, and the present-day machines operating at low energy, are electrostatic. The maximum energy attainable by such accelerators is thus limited by the technology of producing and maintaining extremely high voltages (some millions of volts). This limitation was removed by the invention of the cyclotron by E. O. Lawrence. In a cyclotron, the accelerating voltage is supplied by a high-frequency power source, and the same gap is traversed over and over again by confining the particles in a spiral orbit through the use of a large electromagnet. The geometry is arranged so that synchronism is achieved between the time that the particles transit the gap and the crest of the alternating voltage. A similar idea was also proposed to acelerate particles in a straight line, first by Gustav Ising in Sweden in 1924, but it was not applied to high-energy acceleration until after World War II.

The cyclotron had its limits, in that synchronism between the particle transit and the alternating voltage could not be maintained once particle speeds approached that of light. In that case, relativity limits the particle speed, while the radii of the particle orbits in a magnet continue to increase with increasing particle energy. In addition to achieving particle synchronism with the accelerating voltage, it is also necessary to focus the particle orbits so that, in the long trajectory from start to target, only a small fraction of the particles are lost. The struggle to achieve both synchronism with an alternating field and particle focusing at the same time employed several ingenious ideas; the simple cyclotron does not permit this at particle energies beyond about 20 MeV. One good idea was proposed by L. H. Thomas in 1938, but it was then understood by only few physicists. Thomas proposed an azimuthal variation in the field of the magnet of circular symmetry used by Lawrence in order to achieve focusing and synchronism simultaneously. Thomas's idea was not applied extensively until after the war and then only to machines of moderate energy.

A second idea, which led to much higher energies, was independently conceived in 1945 by Vladimir I. Vexler in the Soviet Union and Edwin M. McMillan in the United States. Vexler and McMillan showed that, if particles were accelerated not at the crest of an alternating field but rather during either the rising or falling part of that field, then dynamic conditions could be achieved in which the phase of acceleration of the particles would automatically be maintained. Those particles that are not accelerated by the ideal amount will arrive either early or late. Depending on the geometrical configuration and the energy, stable acceleration is achieved by an appropriate choice of transit time across the accelerating gap during either rising or falling voltages. Particle focusing can then be attained in a manner less tightly related to the conditions of acceleration. This "phase-stability" principle led to a variety of machines, including the synchrocyclotrons, electron synchrotrons, and proton synchrotrons of the post–World War II era.

The next essential technological advance was the invention of the "strong focusing" principle in 1950 by Nicholas Cristofilos in Greece, and then independently in 1952 by Courant, Snyder, and Livingston in the United States. This principle made it possible to focus particle beams with magnets of much smaller aperture and led, therefore, to a major reduction in the cost of particle accelerators. The combination of this principle with the phase-stability principle provides the basis of all large circular accelerators of today.

As these techniques approach their limits, attention has begun to shift from beams striking stationary targets, which made possible such experiments as the scattering measurements I described earlier, to colliding-beam devices, or colliders, as they are called today. What counts in high-energy collisions in the *energy of collision* observed in that frame of reference in which the total center of mass of the colliding particles is at rest. When a beam of particles strikes a stationary target, at least half of the energy of the incident particle is used to move the center of mass of the two colliding particles ahead, while only the remaining fraction of the energy is available for the collision itself. According to the laws of relativistic collisions, this useful fraction decreases continuously as the energy of the incident beam becomes larger. For example, the energy of collision of a 500 GeV proton striking a stationary target is only 31 GeV. If, on the other hand, two particles, each with an energy of 500 GeV, undergo a head-on collision, then the total collision energy available is 1000 GeV.

This enormous advantage in collision energy is somewhat offset by the fact that the density of practical particle beams is very much lower than that of ordinary matter. Thus, while colliding beams produce dramatic gains in collision *energy,* they lead to large losses in collision *rate*. One cannot have everything, however, and the loss in collision rate can, to some extent, be compensated for by surrounding the points of collision with detectors sufficiently large to catch almost all of the fragments from the events that do occur.

I have digressed to this brief outline of the evolution of accelerating devices during the past 50 years in order to emphasize that, while our studies of the fundamental nature of matter have focused on structures of smaller and smaller size, the actual rate of progress in this field has been determined by the much more mundane matter of successive accelerator technologies. However, as with all exponential growth patterns, the dramatic evolution shown in Figure 15 must sooner or later slow down.

Among the colliding beam devices, the most productive have been electron–positron colliders. An example of such a device was shown in Figure 14. Here, electrons and positrons counterrotate in a ring, and the orbits of these particles are confined by a group of electromagnets. The more powerful recent electron–positron storage rings all use strong focusing magnets; in addition, large radiofrequency power systems are required to compensate for the loss of energy caused by the electromagnetic radiation emitted by the electrons as they traverse their circular orbits.

The advantage of carrying out high-energy physics experiments with electron–positron collisions is the inherent simplicity of the annihilation process. As was indicated in Figure 13, the electrons and positrons indeed annihilate, resulting in what physicists call a virtual photon, which describes a state of pure electromagnetic energy. This electromagnetic energy in turn can rematerialize into any combination of particles that conserves both energy and certain of the symmetry characteristics of the initial collision. Since, unlike collisions in which protons strike material targets, the initial particles have completely disappeared, the final state can be particularly simple and therefore relatively easy to analyze. In particular, the final state can be pairs of particles and their antiparticles, whatever the nature of these may be. These, in retrospect simple, conditions led to the spectacular discoveries of November 1974, aptly called the November Revolution.

The November Revolution

The November Revolution began with the simultaneous publication of two experimental results. The work of Samuel Ting and collaborators at the Brookhaven National Laboratory demonstrated that pairs of electrons produced from a beryllium target bombarded by protons with an energy of 25 GeV in the Alternating Gradient Synchrotron exhibited a peculiar distribution. The correlation in angle and energy of the electron pairs can be expressed as an effective mass of a conjectured object that, when disintegrating into pairs of electrons, would give rise to the observed distribution. The distribution in effective mass observed in the Brookhaven ex-

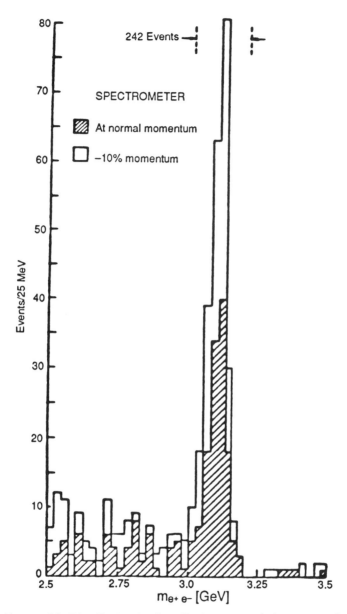

FIGURE 16. *Distribution in the effective mass of electrons and positrons produced at the Alternating Gradient Synchrotron at Brookhaven National Laboratory through impact of a high-energy proton beam on a beryllium target, as observed by Sam Ting and collaborators.*

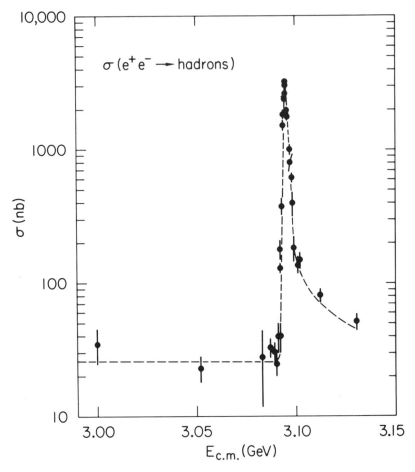

FIGURE 17. *Cross section for the annihilation of high-energy electrons and positrons resulting in the formation of hadrons, plotted as a function of the collision energy of the electrons and positrons, as observed by Burt Richter and collaborators at the SPEAR storage ring at SLAC.*

periments exhibited a sharp peak, as shown in Figure 16, located at an energy of 3.1 GeV. At the same time, experiments carried out by Burton Richter and collaborators with the SPEAR storage ring at SLAC showed that the yield of many kinds of particles—including electrons, muons, pi mesons, neutrons, and protons—showed a sharp peak when the total energy of collision of electrons and positrons was near the same energy: 3.1 GeV. Figure 17 shows these observations.

As future developments indicated, the storage-ring methods proved to be more powerful for investigating these new phenomena, and I will therefore restrict the following discussion to the results obtained with electron–positron storage rings subsequent to the original discoveries.

What made these discoveries a "revolution"? There was little expectation that energy distributions of any kind at these high collision energies could exhibit such a narrow-peaked structure. The reason for this assumption—in retrospect naive—is quite simple: If new particles or objects are formed at these high energies, then there is plenty of energy available for them to decay rapidly into many of the lighter particles that are already known. The larger the energy available for such disintegrations, the shorter the life of the parent particle. And the shorter the lifetime of a new particle, the less well-defined its mass or energy would have to be, according to the energy–time formulation of the uncertainty principle, which is usually written as $\Delta E\,\Delta t \cong h/2\pi$. Thus the only way the sharpness of these peaks could be explained is by postulating a new "selection rule"—that is, some new physical principle that inhibits the rate of disintegration. In turn, a new selection rule must be based on the existence of a new "quantum number" or characteristic property of the particle in question. In short, the November Revolution signalled the discovery of a new kind of fundamental object in nature.

Whenever a label or quantum number changes in some way, then, depending on the dynamics in question, the rates of processes can be inhibited; if nothing changes, then only the available disintegration energy and the masses of the particles participating in the process determine how fast reactions will "go." Starting from this hypothesis, consensus developed rapidly that the new peaks at 3.1 billion electron volts observed at Brookhaven and SLAC must be states of a new kind of quarkonium—that is, objects composed of a new quark and its antiquark. This quark is a new member of the quark family, beyond the three quarks originally postulated by Zweig and Gell-Mann in 1964 to account for the multiplicity of particles then known. The existence of a fourth quark, dubbed the "charmed" quark, had been suspected before, but these new discoveries not only confirmed its existence but, at the same time, as discussed above, provided a new "laboratory" in which the interaction between these new quarks could be investigated.

Charmonium

These discoveries initiated an immediate hunt for the full spectroscopy of this new "charmonium," in complete analogy with the spectroscopy of positronium that I outlined before. This hunt, using electron–positron storage rings worldwide, has proven eminently successful. I shall not describe here the work through which the various spectroscopic states of charmonium were discovered sequentially. There are still some outstanding gaps in that spectrum, and work is in progress to fill these. Some higher-energy states belong to the same family as those of the state (3.1 billion electron volts) originally discovered, and they were found simply by raising the collision energy of electrons and positrons in the storage ring and observing further peaks. Charmonium states of different quantum numbers were located by observing the products of transitions from the directly produced states to the new states. Figure 18 shows this total spectrum as we see it today; and this figure should be compared with Figure 12, which gave the corresponding spectrum for positronium.

Note that the difference in energy scales between these two spectroscopies is a factor of about 100 million, yet the fundamental arrangement of the lines in the two spectra is strikingly similar. This surely indicates strongly that the basic rules of quantum mechanics, which determine the existence of such discrete states and their energy, are the same whether one deals with electron volts, millions of electron volts, or billions of electron volts. Note that the proportions in spacing among the lines are, however, greatly different. This is not unexpected, since the positronium spectrum is solely governed by the laws of electricity and magnetism, while the charmonium spectrum is governed by the not-as-yet fully understood laws of quantum chromodynamics; this is the name given to the rules that govern the forces between quarks as they exchange gluons with one another. Naturally, many investigations have been carried out by theorists attempting to fit the new spectra under various assumptions, drawing on the analogy with positronium. One exploits this analogy by comparing the simple inverse-square law of the electromagnetic interaction that governs the positronium spectrum with the appropriate radial dependence of the force that can account for the charmonium spectrum. This quest has thus far been only partially successful.

Beyond Charmonium

The spectroscopy of charmonium, analogous to that of positronium, is not the only result from electron–positron annihilation at high energies. His-

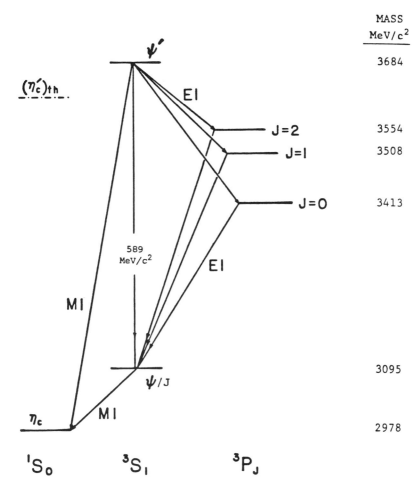

FIGURE 18. *The energy level spectrum of charmonium. Three classes of spectroscopic terms* (1_{S_o}, 3_{S_o}, 3_{P_j}) *are shown using the usual spectroscopic notation for spin multiplicity, orbital angular momentum, and spin angular momentum. The observed masses in millions of electron volts divided by the square of the velocity of light (MeV/c^2) are shown on the right-hand side of the diagram, and the observed transitions are designated by arrows. Note that the spectral lines in classification are the same as shown in Figure 12. However, the scale of the principal transition is different by a factor of about 100 million (589 MeV/c^2 relative to 5.1 eV/c^2). Moreover, the separations among the lines of differing angular momenta are very much larger than those for positronium.*

tory was again to repeat itself, as accelerators and electron–positron storage rings shifted to higher energies. At Fermilab, a fifth member of the
quark family, the b quark (b for bottom) was discovered. Subsequently,
the combination of the b and anti-b quarks—"bottomonium"—was observed in the form of a peak in the yield of various particles as a function
of electron–positron annihilation energy at the storage ring DORIS in
Hamburg, similar to earlier SPEAR observations for charmonium. Some
of the more detailed spectroscopy of this object began to unfold at the
Cornell storage ring CESR in the United States. Although these spectra
are not as yet as complete as those from the charmed quark states, it is
clear that we are again seeing the same basic quantum states as those in
positronium. Interestingly, the utility of the spectroscopy as a laboratory
to explore the forces acting between quarks increases as the masses of the
quark pairs under study increase. The reason is, the heavier the quarks
are, the slower they move in the "quarkonium" combination, and therefore the simpler the theoretical description becomes, because the effects
of relativity need not be included.

Electron–positron annihilation gives a very direct signature on the
number of quarks that contribute to hadronic matter as we know it. By
"hadronic matter," we mean the totality of those particles in nature (including the neutron, the proton and many others) that interact through the
strong force—that is, the force we believe to be responsible for binding
the neutron and proton in the nucleus. Look again at Figure 13, and note
that the initial interaction at the first vertex, in which the virtual photon
first forms a pair of quarks, is purely electromagnetic. In consequence,
the theoretical calculations giving the probability of these reactions is
simply proportional to the sum of the squares of the electric charges of
each possible quark that could contribute to the process being investigated. As the energy of electron–positron annihilation increases, presumably more and more kinds of quark pairs can be produced, depending on
the masses of the quarks and the energies of the annihilations.

Figure 19 describes the current picture of the probability of annihilation of electrons and positrons as a function of energy. What we see are
essentially two distinct plateaus separated by a transition region. The
arithmetic shows that the first plateau corresponds, at least fairly closely,
to three "flavors" of quarks, while the second plateau corresponds to the
addition of a fourth, the charmed, quark flavor. Thus this general picture
confirms the interpretation of the peaks, which gave the initial incentive
to the November Revolution, as being pairs of charmed quarks. Again, as

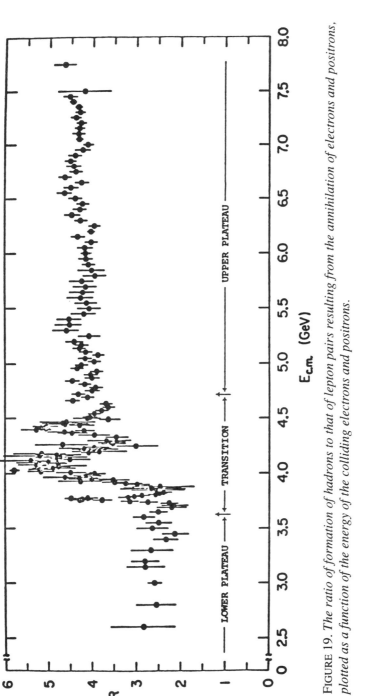

FIGURE 19. *The ratio of formation of hadrons to that of lepton pairs resulting from the annihilation of electrons and positrons, plotted as a function of the energy of the colliding electrons and positrons.*

in some of the previous examples, detailed quantitative comparison of this simple picture with the data shows some discrepancies; and again we can learn, from reconciling these differences, more about the detailed dynamics of the pertinent processes. Thus, just as the deviation from the simple scaling relationship has taught us something about the strength of the interactions among the quarks in the proton, so too can we learn, from the discrepancies between the exact numerical predictions concerning annihilation cross sections and the experimental results, how quarks interact with one another once they are formed.

The story I have described is, of course, only a small fragment of the evolution of our understanding of particle substructure during the century. What I hope, however, is that this sketch makes clear that we are witnessing an amazing contrast between the surprising new discoveries paced by the evolution of technology, on the one hand, and a consistency of general approach and motivation and of overriding physical principles, on the other. It is this contrast between sameness and change that has made this period one of the most exciting epochs in physical discovery.

Basic Research: Curse or Blessing?

W e are gathered here in order to celebrate the twenty-fifth anniversary of DESY. I am particularly pleased to participate in this celebration for several reasons. First, DESY is a sister institute to the Stanford Linear Accelerator Center—SLAC, for short—which is a part of Stanford University in California, which has been my place of work for the last 34 years. SLAC and DESY work on similar topics—we both operate electron accelerators and electron–positron storage rings. Our institutes are of similar size, and we have about the same number of staff members. We both are now starting to build colliding-beam machines using new principles.

Second, I am pleased to be here in order to congratulate DESY for the great progress in science that has been set in motion from here. But this visit is also a sentimental occasion for me. I lived in Hamburg as a child, from 1919 until 1934. I was educated at the Gelehrtenschule des Johanneums, and my father taught at the University of Hamburg.

DESY's work is a part of the worldwide effort whose goal is to discover the secrets of nature. We all know that, since the time of the ancient Greeks, it has been a human aspiration to know from which building blocks the world is constructed. The smallest part of matter that in itself is still "the same" was called by the Greeks "atom" or "indivisible." Today we know that the atom has approximately a diameter of 10^{-8} centime-

A talk given at Deutsches Elektronen Synchrotron (DESY) September 24, 1984.

ter; this means that a cut through a human hair crosses approximately 1 million atoms. Today, physicists have advanced to the unimaginably small distance of 10^{-16} centimeter = 10^{-8} centimeter × 10^{-8}. DESY has contributed a great deal to this progress. Today we can resolve items that are smaller than the size of an atom by the same factor by which the atom is smaller than 1 centimeter.

Modern science attacks simultaneously the "limits of the small" through elementary-particle physics and the "limits of the large" through cosmology and space exploration. It is particularly worth emphasizing that, during the last few decades, these two have been mutually reinforcing. In order to understand the phenomena in outer space, one needs the results of particle physics. Both fields of research—space exploration and particle physics —require large apparatus. In order to explore small dimensions, one must give very high energy to the particle probes; this follows from the uncertainty principle of Heisenberg, which states that a small dimension can be determined only if it is probed with particles of very high energy. For this reason, the terms "high-energy physics" and "particle physics" designate the same topic of research. As the energy of our accelerators and storage rings grows, we can explore smaller and smaller dimensions in proportion. But this is not all. Not only is the physics of the highest energies and the physics of the smallest particle the same thing, but so does the exploration of the cosmos rest more and more on the results of particle physics. The rules that determine the generation of energy in the far regions of space derive from results that have originated in such laboratories as DESY. We see, therefore, that the results of particle physics have assumed the role of the great unifier of different parts of natural science. It is not very surprising that the improved understanding of the fundamental building blocks of matter and of the forces that link these building blocks leads also to improved understanding of many other phenomena in nature.

We see that particle physics is a great unifier of different topics of science. However, particle physics is also a unifier of people. Physicists and other scientists throughout the world who work on this subject know one another and share a mutual respect. There are no secrets in particle physics. All new results are published promptly. At meetings of physicists dealing with this topic, one hears all languages. Technical lectures in many countries are generally delivered in only one language—English— independent of the nationality of the speaker or the locale of the lecture. I am delivering this talk in bad German, since our auditors are not all tech-

The World of High-Energy Accelerators

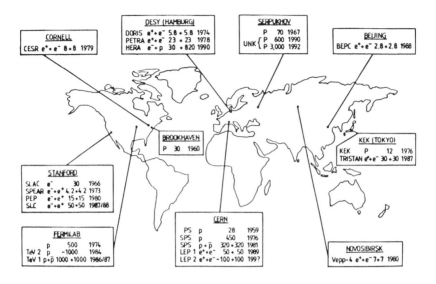

FIGURE 1. *Location of machines that accelerate particles to energies in excess of 1 billion electron volts.*

nical people! It is particularly important not only that the role of particle physics is recognized by specialists but that all citizens who share an interest in this subject have the opportunity to understand it. After all, we are all paying the bill for the costs incurred in carrying on work on this subject.

Laboratories that are active in this field exist in almost all countries of the world. The first illustration (Figure 1) indicates, on a map of the world, the location of each of the machines, either now in operation or still under construction, that accelerate atomic particles to energies in excess of 1 billion electron volts. There is no such thing as "the best machine in the world" for particle physics. These laboratories operate installations of many kinds: accelerators or storage rings, electron or proton machines, weak beams or intensive beams. Each machine is particularly effective in exploring a special topic; the ideal machine for all purposes has not been invented and is indeed not inventable. I am emphasizing these facts in order to convince you that international collaboration in this field not only serves international culture and international peace and understanding but also is necessary for purely scientific reasons.

I would like to cite three current examples of results from many countries that are mutually supporting. In that area of research in which one wishes to understand the structure of the proton at the highest level of detail, HERA here in Hamburg will be, without doubt, the most effective machine. In order to extend the technique of elementary particle collisions (which has been so successful) to the highest energy, the LEP storage ring in Geneva, which is now being built at CERN, is without question the best tool. In order to produce the recently discovered Z^0 particles relatively cheaply, and in order to demonstrate simultaneously a new technology for colliders, the SLC (Stanford Linear Collider), which we are building in California, is the most suitable device.

The goal of particle physics is basic research on the nature of matter. Historically, we know that basic research has always led to applications. We do not know, of course, in which areas of human endeavor particle physics will find application—nor do we know, in fact, whether the results of the physics of elementary particles will find *any* application. But we must remind ourselves that nuclear physicists in the 1930s did not think about practical applications. The great British physicist Ernest Rutherford, the discoverer of the nucleus, declared, "Who says that nuclear energy can be practically used speaks moonshine." But today we have radioactive medicines from nuclear reactors, nuclear power plants, and even nuclear bombs. The contribution of nuclear physics to medicine is without question a boon to mankind; I believe personally that this is also the case for nuclear power plants, although this is a controversial matter. Nuclear weapons, however, are a burden that humanity now must carry. But, in spite of this—or maybe even because of it—there has not been a world war for 39 years, which is much longer than the short period between World War I and World War II.

Are the results of nuclear physics a curse or a blessing? Are the future results of particle physics generated by DESY and other institutes for fundamental research a curse or a blessing? There are few questions that are more important for the future of the world than these.

It is very simple for the citizen who is lamenting the threat of nuclear weapons or the degradation of the environment to place the responsibility on science. This impulse is simple but wrong. The laws of nature exist irrespective of whether we discover them today, tomorrow, or the day after tomorrow. For example, we know today that the first nuclear reactor was not the one produced by Fermi and his collaborators in Chicago during the war in 1943; a nuclear fission explosion of natural origin occurred in Gabon on the west coast of Africa roughly 2 billion years ago. This fact was discovered 12 years ago by analyzing the ore produced in a uranium mine at that location. Attempting to deny knowledge to humanity is no answer; nature's answers will reveal themselves sooner or later.

The transition from basic science to useful technical application and finally to mass production constitutes a long chain. The links of this chain are

- experimental and theoretical basic research;
- synthesis of results;
- invention of an application;
- development of a product;
- tests on a prototype;
- evaluation of the tests; and
- production and distribution of the product.

Only the first link in this chain is dedicated to the discovery of the properties of nature—these properties exist without human participation at any account; they are a part of our environment whether we explore them or understand them or not. The further links of this chain, starting with a proposed application and ending with production and distribution, are under the control of human beings: if people do not develop a product, then that product does not exist. The great question is, therefore, Are organized human institutions and individual human beings capable of making wise decisions about the application of basic knowledge? My conviction is that humankind and its institutions make better and more certain decisions concerning the application of a natural law when they understand the basic substance of such a law of nature.

It is very easy to complain that certain influences of technology have complicated life. One must, however, compare such aggravations with the catastrophe that would arise if one attempted to feed, clothe, house, transport, and protect the existing and future population of the world without the aid of further technical developments.

Technology has eased the exchange of information among people of all countries. A result is that secrets are more difficult to keep secret, both in private and in public life. Although this may cause discomfort in private affairs, one must admit that openness in international affairs can be an important step towards peace. The satellites that now orbit the earth have eliminated many of the secrecy barriers among nations. The control of weapons in the world is imaginable because the satellites and radars can accomplish their observations above and through the Iron Curtain. The result of such technical developments is that we now live in a much more open world than in the previous century.

The future of civilization is to a large extent dependent on progress in technology. In turn, the progress of technology depends on progress in basic research. Thus, human society has no alternative to carrying out intensive scientific research. The only question is, How intensive?

Basic research requires money and work. When the economy is in difficulty, one often hears the question, "Yes, but can we afford basic research?" The answer is that, without basic research or technology, the economy cannot be healthy. This would be a tree without roots, which cannot thrive for any length of time. If money is short, it often appears to be the simplest solution to defer everything that serves only the long-term future. However, whoever chooses such a course mortgages the future.

Figure 2 shows schematically how costs are distributed along the development chain, from basic research to distribution. One sees from Figure 2 that the costs of basic research are only a tiny fraction of the total cost of development. Yet, without basic research, the technical end product would not even exist; basic research is the seed for all further technical development. Let me give an example. The principle of the linear accelerator was discovered before World War II and was reduced to practice at a research cost of about DM 1 million. This development has led to an industry that develops linear accelerators for cancer therapy at hospitals. The total price of linear accelerators developed for this purpose to date is roughly DM 10 billion. The first link of our development chain—that is, theoretical and experimental basic research—will surely, although at some future time, uncover new truths about nature. Therefore, the answer to the question—"basic research: curse or blessing?"—is determined in part by moral and political concerns of society.

Can we direct the results of research into constructive directions and avoid simultaneously those applications that might have destructive consequences?

Nowhere is this question more important than in military policy. Nature has given us the means to destroy all civilization on earth. I am saying here "nature" and not "science," since I am persuaded that the secrets of nature will exhibit themselves openly sooner or later. The source of energy for nuclear weapons is the same source that furnishes fuel to the sun

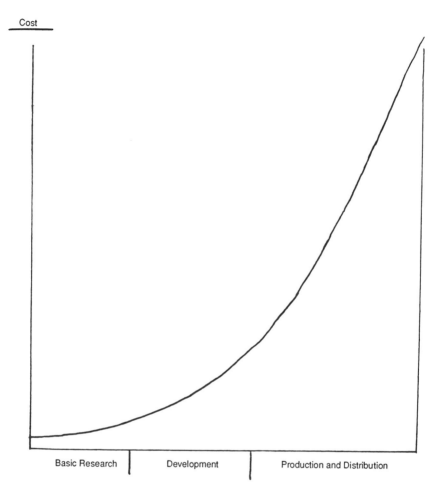

FIGURE 2. *Distribution of costs along the development chain.*

and the stars.

At this time, there exist roughly 50,000 nuclear weapons in the world, roughly evenly divided between NATO and the nations of the Warsaw Pact. This number is much too large to be applicable in any way in case of war without threatening the future of civilization. The number of nuclear weapons is much larger than can be justified for deterrence of war. We know this: the Americans know this, and the Russians know

The Number of Strategic Warheads

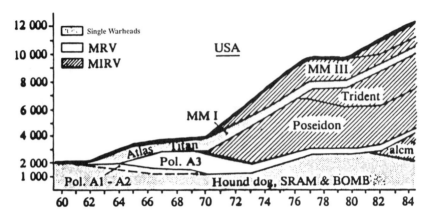

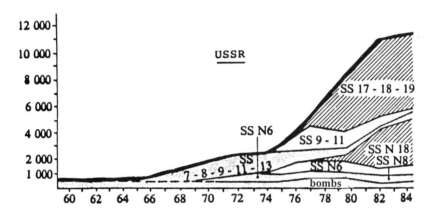

FIGURE 3. *The number of strategic warheads.*

this. Nevertheless, the number of nuclear weapons has risen to this un-imaginable figure. How did it happen? Much has been written on this subject. The principal reason is that states, whenever they acquire new weapons, justify the need—above all, politically and not scientifically. Fi-gure 3 shows as an example the mutual competition in increasing the number of nuclear weapons that can be delivered with long-range strate-gic weapons systems between the Americans and the Soviets. Such a competition can only be justified when weapons have become *symbols* of strength without reference to the *reality* of their military usefulness. We scientists have the earnest duty to draw the attention of politicians repeatedly to the physical realities of nuclear weapons. If, in the political process, the scientific reality of weapons is forgotten, then the fateful question, When is enough enough? cannot be answered. In this sense, science can be a blessing, if it succeeds in convincing politicians and the public that the physical realities of nuclear weapons—that is, their destructive power—is so overwhelming that, for the goal of deterrence, only a very small number can be justified. Generally, the actions of politicians are determined by perceptions and not by reality. The scientist attempts to single out the truth in nature—that is, the objective reality.

One can easily find many other examples in which the application of science to technology is either beneficial or damaging; what actually results depends on whether the people in decision-making roles recognize clearly the scientific circumstances and react logically to them, or whether they push the scientific facts into the background.

We must learn that the forces of nature cannot be coerced by the state; we can neither change nor hide the fundamental laws of nature. We know of many historical examples in which a governmental or religious power decided that a law of nature was uncomfortable. In most such cases, the results were tragic. During the Renaissance, Galileo became convinced on the basis of preceding astronomical observations, that the earth orbits around the sun and is not therefore, the center of the world. However, he was forced to recant this opinion under threat of torture and death. In the Soviet Union, the genetic theories of the agronomist Trofim Lysenko, which were based largely on wishful thinking and the desire to be politically correct, were elevated to an edict of the state in order to justify the state's ill-advised agricultural policy. Only after a courageous fight on the part of the Soviet Academy of Sciences was the teaching of genetics as a true science again permitted in the Soviet Union.

In Germany, between 1933 and 1945, it was rigidly forbidden to teach Einstein's theory of relativity. Today, the theory of relativity forms the basis for many fields of technology. Without the theory of relativity, there would be no high-powered electron tubes, no intense x rays, no high-energy accelerators, and therefore no DESY. And the criminal decision by the state, in contradiction to science, to attribute to one human group a fundamental racial superiority led to the greatest tragedy for the German people, for their victims, and for humanity.

Today, humanity trembles before the threatening might of many thousands of atomic weapons. Nevertheless, the president of the United States has proposed a new defensive umbrella that will, he asserts, make the existing nuclear weapons "obsolete and impotent." This proposal was made despite the fact that the immense power of today's nuclear weapons has given the offensive power an enormous advantage against defensive measures. What science has invented cannot be uninvented through official decree.

The situation is similar in respect to environmental protection. It is senseless to assert that science and technology are the basic causes of the pollution of nature; one should look for the causes in the unwise decisions of human society. We need science, both basic research and the development of specific remedies, in order to clean up our environment.

With all of these examples, I am trying to demonstrate that, without basic research that leads to a better understanding of nature, the growing number of people cannot be accommodated on this planet without great suffering and catastrophe. Basic research is therefore absolutely necessary for the future of civilization. With science, it is at least possible that we can live in harmony with the natural world. However, this can become reality only when our citizens listen critically to scientific advice. In a fundamental sense, science is not a curse, but it can become a blessing only if it is treated and explored carefully. Unfortunately, many politicians and many parts of society attempt to exploit only the great possibilities of science for industrial application, and they hide simultaneously the results of the same science indicating limits of growth and the necessity of restricting military might. The truths of the laws of nature are inseparable: We cannot pick out those laws of nature with which we feel comfortable and ignore those that make us uncomfortable.

In the foregoing remarks, I have implied that science is a two-edged sword. I have drawn the conclusion that research, including basic research, is a necessity of modern society and that it makes possible the improvement of the quality of life throughout the whole world—but it does not assure such improvement. DESY can be proud to have played a very important role in this undertaking. This statement applies not only to the past but also to the future. The new machine HERA will provide a unique opportunity for DESY to find new fundamental results in particle physics. HERA is the only collider in the world in which electrons and protons collide.

With this apparatus, the structure of protons can be explored with the

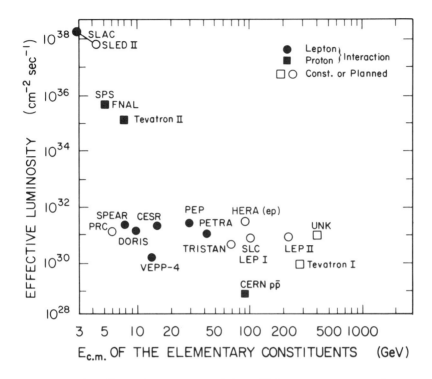

FIGURE 4. *The effective luminosity and the collision energy among elementary constituents (quarks, gluons or leptons) in their center-of-mass system, of existing or future colliding beam machines.*

highest precision, since the structure and interaction of electrons are well known. It is interesting that the three large European colliders—HERA, the CERN proton–antiproton adaptation of the super proton synchrotron (SPS), and the first step of the LEP machine at CERN—all lead to the same collision energies, if one measures the energy with which the fundamental building blocks (leptons, quarks) collide. This is illustrated in Figure 4.

The future of laboratories like DESY and my own institution, SLAC, depends on continuing renewal. When DESY was founded 25 years ago, no one could imagine the future with DORIS, PETRA, and HERA. As illustrated in Figure 5, the progress of high-energy physics is paced by inventions of new means of acceleration. Each new method of acceleration of elementary particles makes it possible to raise the attainable energy limit. Yet each one method can only lead to a certain degree of progress

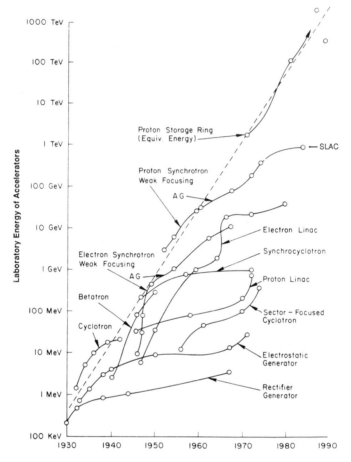

FIGURE 5. *Laboratory energy (or equivalent laboratory energy in case of colliders) as function of time.*

before it becomes too expensive; then a new technology must be invented and introduced. DESY has contributed to many of these jumps; I believe that DESY will also contribute to the subsequent advances.

I am pointing out these facts to demonstrate the systematics of these developments, which show how basic research leads to the progress of technology, but also how, conversely, that new technology is necessary for the progress of research. Science and technology have a mutually reinforcing symbiotic connection.

I would like to conclude with congratulations to DESY, a laboratory that has played such a large role through a quarter century in this great adventure of our time.

Big and Small Science

The Evolution Toward "Big Science"

I t is not news that the scale of scientific endeavors required to obtain answers at the frontiers of science has increased and continues to increase. Although public concern with this growth has risen dramatically with specific recent proposals, such as the Superconducting Supercollider (SSC) in the United States, the Large Hadron Collider at CERN, and the Human Genome Project, there has in fact not been a sudden onset of big science initiatives but a gradual evolution through many decades. It is characteristic of most scientific endeavors and technical developments that they undergo exponential growth until some mechanism that terminates that pattern interferes. I would like to give some examples of this phenomenon in this talk and also to discuss some of the criticisms that have been levied against big science. In particular, I would like to maintain that the driving force in the evolution of the size of basic science is controlled by scientific and technological developments and that the spirit and motivation of those pursuing such endeavors has not changed.

The primary example of the evolution in size of scientific endeavors is, of course, elementary-particle physics. The study of elementary particles has evolved from what is now known as nuclear physics to high-energy physics in the giga-electron-volt (GeV) region, and concomitant with this has been the growth of accelerator energies. Figures 1 and 2 show the usual Livingston graph, which charts the growth in the energy of accelerators in equivalent laboratory energy since 1930. That quantity still exhibits pure exponential growth, which is nourished by a succession of

A talk given at the University of Rome, September 20, 1988.

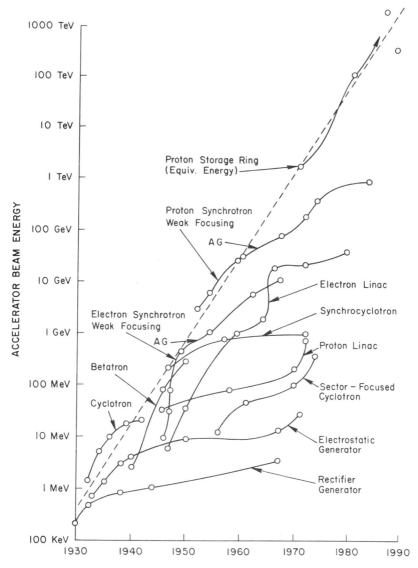

FIGURE 1. *The Livingston chart: growth of accelerator-beam energy with time.*

technologies: as one accelerating technique has reached saturation, new inventions have come to the rescue. There are very similar patterns pertaining to the technology of computers: the processing power of computers is still exhibiting pure exponential growth. The capacity of integrated

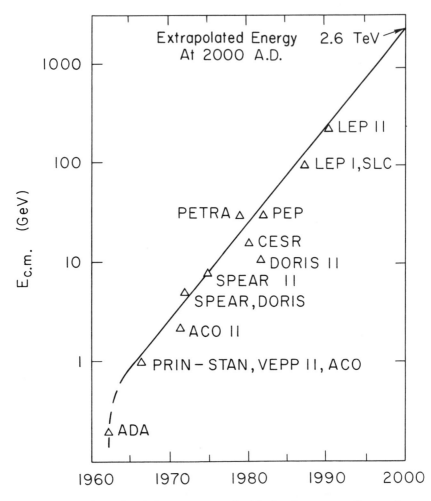

FIGURE 2. *Growth of the energy attainable in the center-of-mass frame* (*E*cm) *of the fundamental constituents involved in the collision. Fundamental constituents are either electrons, protons, quarks, or gluons, which exhibit a "pointlike" structure at the highest energies attainable thus far. This energy is simply the sum of the beam energies of electrons and positrons in electron–positron colliders. In the case of proton colliders, the constituent frame energy is roughly a factor of 10 lower than that of the energy of the protons, since the proton energy is shared among constituent gluons and quarks.*

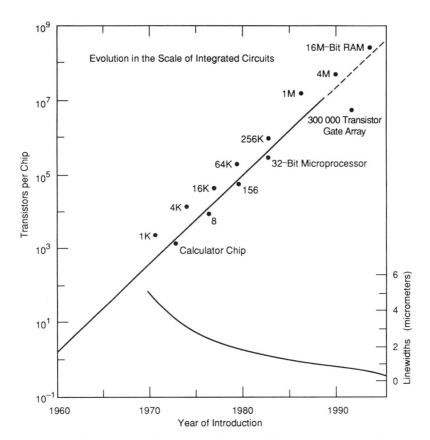

FIGURE 3. *The increase of transistor density of the most pervasive IC types, memory and logic, has been steady and predictable; about every two years the density of active elements doubles on logic chips and quadruples on memory chips. Meanwhile, linewidths have steadily decreased, although recently the slope has flattened, reflecting the greater difficulty of scaling down by tenths of micrometers.*

circuits follows such a pattern, as is shown in Figure 3. Such growth is, of course, only possible if unit costs (that is, the costs per GeV of accelerator particles or the cost per computer throughput or memory unit) decrease. Figure 4 shows this pattern applied to accelerators. However, as is frequently the case, the decrease in unit cost is not as rapid as the increase in desired performance. Therefore, the cost of each new installation or acquisition tends to increase. Of course, the amount of this increase is much

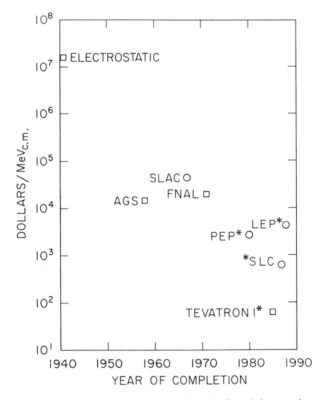

* Based on incremental costs in existing laboratories
and using existing injection systems

FIGURE 4. *Rough estimate of the construction cost (in dollars per unit of attained collision energy in the center-of-mass system) for accelerators and colliders between 1940 and the end of the century. Costs are corrected for inflation.*

less than the increase in the performance parameters. In high-energy physics, the cost per unit of energy has decreased by perhaps five orders of magnitude since the 1930s, while the energy itself has increased by seven orders of magnitude, and the cost of each new installation, therefore, has increased by perhaps a factor of 100.

We do not know, of course, what factors will eventually terminate the exponential growth of collider energies. Such factors could be financial, technical, or social. On the *financial* side, I do not believe that the cost of accelerators, even that of the SSC and the Eloisatron, which are the largest machines now under consideration, are in themselves as yet prohibitive, considering the small fraction of GNP represented by research, and considering the very small ratio of basic to applied research. *Technically*, however, there are serious problems. Even though the cost of proton synchrotrons using superconducting magnets scales linearly with energy, the ratio of the cross section for new high-energy events to the total cross section varies inversely with the square of the energy or mass range under investigation. Thus, the detection and data-analysis problems become extraordinarily difficult with increasing energy.

It is extremely unlikely that new technology—for instance, technology associated with the new high-temperature superconductors—has any bearing on the future of proton synchrotrons. For electron–positron linear colliders, the technical future looks good for some additional steps in energy. However, here the demands on beam quality and on power sources escalate dramatically with energy, and thus we are entering an entirely new realm of precision, instrumentation, and control. At this time, we cannot forecast precisely how far such new technologies can reach, but the difficulties are profound.

One interesting fact is that, in parallel with this dramatic shift in performance of the apparatus of modern science, the time to construct a new facility has hardly changed at all. The building of the first accelerator at CERN, the construction of the SLAC linear accelerator, and the creation of the proton synchrotron at Fermilab all required between 4 and 6 years, not significantly different from the construction time of the early electrostatic generators. The reason for this is similar to the reason why a flea and an elephant jump to about the same height: the ratio of weight to muscle power for both animals is about the same. The organization of building the more recent large machines is designed to permit a very much larger number of people to participate in the construction.

Thus, while we anticipate eventual technical limits, these cannot be identified at this time, and some further steps are clearly possible. But there remain *social* problems. Will these machines become so large and so difficult either to build or to use that the social environment necessary will be incompatible with the fundamental spirit and creativity necessary for basic research? This is the question to which these remarks are addressed.

Thus, during this dramatic period of evolution, we see many similarities with the past, but also many changes. The question is, how has this affected the actual conduct and creativity of big vs. small science. Recently, "big-science bashing" has become popular. For instance, there was a recent article in *Issues in Science and Technology* that attributed the cause of the reported instances of scientific fraud to a decline in morality brought on by big science, notwithstanding the fact that all the much publicized instances of such fraud occurred in medical research—and, in particular, in medical research carried out on a laboratory scale.

The criticism of big science is often motivated simply by the perception that the large cost of big-science facilities competes destructively with the support for small science under the assumption that the total "pot" for the support of all of basic science is constant. Yet evidence from the pattern of science support in past years speaks to the contrary. In parallel with the evolution of the large instruments of science, the total support for basic science has increased; that is, the fraction of the total support of science that has been dedicated to the construction of large scientific instruments has stayed about constant. In other words, at least in the past, the model of big and small science patterned after big and small ships on the same ocean going up and down together has been more applicable than the model that assumes that there is direct competition for a fixed total resource. This may, of course, not be true in the future, and there is no way to predict what will be the relation between the support of big science projects like the SSC and the expectation of support for all of basic science. Generally, intentions appear to be good; for instance, the U.S. administration, in expressing its support for the SSC, also proposed substantial increases for the National Science Foundation, which is the primary agency supporting small science in areas other than medical research.

The Growth of Big Science at Stanford University

My previous remarks have described the broad pattern of growth. Let me first illustrate the evolution from small to big science at my home institution, Stanford University, and then make some comments on the actual conduct of the work.

The largest single installation at Stanford, SLAC, was not the result of a sudden transition to big science but was a step in a pattern of growth that was initiated during the Korean War. At that time, a deliberate decision was made by Stanford Provost Fred Terman in consonance with several faculty members, including those of the Physics Department, that the university should expand its involvement in government-supported scientific and technical activities. A primary motive for doing so at that time was the fear that, during the Korean War, the technical faculty would leave Stanford, following a pattern that had been established during World War II.

The evolution of linear accelerators at Stanford was initiated by Bill Hansen before World War II. He invented the resonant cavity specifically with the motivation to produce high voltage for accelerating charged particles. His invention was initially sidetracked to its application to microwave technology—in particular, microwave power tubes. In turn, that invention, especially the development of the klystron tube, aided linear accelerators. Construction of a sequence of electron linacs was initiated by Bill Hansen and continued under the direction of Edward Ginzton.

In parallel with the development of the electron linear accelerator at Stanford, Luis Alvarez and his colleagues at the University of California, Berkeley, constructed a pioneering proton linear accelerator with a maximum energy of 32 MeV. In the late 1940s that proton accelerator appeared to be a leap forward toward bigness, while, because of a higher operating frequency, the electron linear accelerators remained small. Alvarez and his associates were photographed sitting on the vacuum tank of their machine; upon learning of this photograph, Hansen demonstrated that he could carry his machine on his shoulder. The comparison is shown in Figure 5. Thus we have one example of big versus small science: You can sit on big science, but you can carry small science. Needless to say, the situation has not always remained this way. Figure 6 is a photograph of the Stanford Linear Accelerator Center, which stems from the evolution of Hansen's initiatives. It is clearly not portable!

Due to the pressure of developing klystrons for applied purposes, two independent laboratories, the Microwave Laboratory and the High Energy Physics Laboratory (HEPL), were established as separate units sharing a common administrative structure. The two accelerators called the Mark II and the Mark III, following Hansen's initial pilot machine called the Mark I, were not "small" science. Budgets were in the multimillion-dollar-per-year range (when a million dollars was still a million dollars). Crews of professional operating technicians were necessary, and the beam of the Mark III accelerator was shared among a multitude of users. With few exceptions, these users were, however, "in-house"—that is, members of the Stanford faculty, students, and staff. Most of the work

FIGURE 5. *The top picture shows the tank intended to house Luis Alvarez's 40-foot proton linear accelerator at what was then the University of California Radiation Laboratory at Berkeley. The collaborators in accelerator construction are sitting on the tank. Not to be outdone, William W. Hansen of Stanford University had himself photographed with his three collaborators carrying the Mark I Linear Accelerator at Stanford. He intended to demonstrate that this was "small science," in contrast to the upper picture, which was "big science."*

was done collaboratively, but these groups were small by today's standards in high-energy physics, and, generally, a member of the Stanford faculty directed each collaboration. The laboratory was the responsibility of a director who was not necessarily a member of the regular faculty.

Mark III was a great success, leading to many important discoveries in particle physics, including the Nobel Prize-winning work on the size and shape of the proton and neutron. This lent encouragement to studies that eventually culminated in a proposal for the construction of the 2-mile linear accelerator, a laboratory that came to be called SLAC, after some search for alternative acronyms. A few relevant points in connection with the birth of SLAC are worth mentioning:

FIGURE 6. *Aerial photograph of the 2-mile Stanford Linear Accelerator at Stanford University, which is "big science," but which is an outgrowth of the small-science efforts of Bill Hansen. Shown are the linear accelerator itself, its target areas (end stations), the central laboratory complex, and the associated aboveground detector experimental halls, in which colliding-beam experiments in storage rings and other colliders are carried out.*

1. SLAC was the largest accelerator proposal at the time, and it remained the most expensive single installation in high-energy physics until the construction of the National Accelerator Laboratory, now Fermilab.

2. The community of physicists was not enthusiastic about the creation of SLAC. At the time of its creation, the mainstream of high-energy physics used protons as the bombarding particle, and wisdom of switching to the use of an electron beam delivered in short,

widely spaced pulses was widely doubted. Yet, at the time of the
SLAC proposal, science budgets were growing at roughly 15 percent per year, so there was no feeling of competition or threat from
the new facility to the programs of other sciences. The type of de-
visive debate that we are seeing today with respect to the SSC was
absent.

3.   HEPL was a facility of Stanford University, although it was operated outside the regular departmental framework. When SLAC
was initiated, there was a debate within the physics community at
Stanford about whether SLAC should be operated like HEPL, described somewhat simplistically as an arm of the regular Stanford
Physics Department, or should become more national in character. The decision was made and accepted by the government that
SLAC should be a "national facility," meaning not quite a national
laboratory. The meaning of this term evolved to signify that SLAC
should be available equitably to scientists throughout the world,
with judgment on priorities to be based on scientific merit and demonstrable ability to carry out the proposed program.

A committee structure was established, including a Scientific Policy
Committee to advise the president of the university, designed primarily to
safeguard the rights of the outside users. Yet the line responsibility for
running SLAC was strictly within Stanford. It devolved from the president, to the director, to the line organization of the laboratory. This pattern is unique. Other high-energy physics laboratories in this country and
abroad have become what Leon Lederman, the current director of Fermilab, has characterized as "truly national laboratories." At the same time,
other large accelerator facilities, notable the Cornell accelerator, have remained proprietary. With the benefit of hindsight, I consider much of the
debate about national-laboratory, national-facility, and even proprietary
campus machines to be a "tempest in a teapot," as long as we are dealing
with installations that require a major professional presence for their design, maintenance, and operation. As a practical matter, the social interactions between the professionals, the inside users, and the outside participants are much more controlled by the state of the science and technical
circumstances than they are by the formal arrangements for allocating
use. The decision-making process, the review mechanisms by outside
committees, and the relationship to government oversight are all remark-
ably similar.

Big Science and Academe

The advent of big science has raised a series or profound questions with respect to the relation between academic structure and the management of big science laboratories. Dealing with these issues requires administrative inventiveness. Blind insistence on preserving academic practices that are no longer applicable is counterproductive, yet exclusive emphasis on such matters as administrative efficiency, financial accountability, and so on, in disregard of academic values is at least as destructive.

Academic institutions have become responsible for big science laboratories as managers and operators of the facilities, and also through the participation of their faculties, students, and staff in the research work. It is the second role that challenges traditional practices.

Work in big science requires collaborative efforts by large groups. However, the dynamics of such efforts is frequently misunderstood. Indeed, the construction of an accelerator is a major collaborative effort, and, in the United States, the potential users frequently get involved in many phases of such construction. Physics users participate in workshops determining the fundamental design parameters of the machine, they maintain contact with the builders to make sure provisions for future experiments are adequate, and they frequently participate directly in the commissioning of the machine, both to accelerate its completion and to become acquainted with its detailed operating characteristics.

Design and construction of particle detectors also tend to be major cooperative enterprises with frequently more than 100 physicists from many institutions participating in recent times. Again, as many physicists participate in data taking with such a facility. Such data taking usually extends over all hours of day and night, and the software effort associated with data acquisition is enormous.

Note that, in "big science" of this type, many more data are recorded than are required for the primary experimental goals. The real scientific exploitation of the data commences when "mining the tapes." It is during this later phase of data analysis that individual ingenuity becomes paramount. Frequently, important discoveries in physics have been made years after the primary data have been taken. The primary example of this at SLAC was the discovery of the tau lepton. Here a large collaboration of physicists built the colliding-beam machine SPEAR and then constructed the Mark I detector. This led, among other triumphs, to the discovery of the Ψ/J particle and the elucidation of the spectroscopy of psions, which are composed of the charmed quark and its antiparticle. However, about two years after these discoveries, Martin Perl and his students carefully massaged the data for an excess of coincidences between detected electrons and muons. Such coincidences can only be explained if initially a pair of particles is formed that can independently decay into electrons and muons. This conjecture was converted into what became irrefutable evidence for the existence of a third generation of leptons known as tau particles.

It should be noted that, throughout this process of "big science at SLAC," the motivation of the participating academic physicists remained exactly the same as it had been in smaller endeavors—that is, to uncover new basic facts of nature. However, because of the ways in which the work is carried out, there are indeed problems in academic practices that must be resolved. Among these are:

- *Recognition of Individual Contributors.* Since most but not all of the participation by physicists is in the form of group activities, individual contributions cannot easily be traced through the publication record. Therefore, during faculty searches and promotion inquiries, reference must be made to personal contacts, with bibliographical data being used for backup only.
- *The Tenure Clock.* Attainment of a truly significant scientific result may require a time span not under the control of a young physicist participating in the work. At the same time, the "up or out" practices on the academic ladder require documentable accomplishment during a fixed time. The question is whether the "tenure clock" can be stopped while the individual participates, say, in the construction of a major detector.
- *Thesis Standards.* Ph.D. theses are expected to be "independent pieces of work." As a practical matter, this standard is frequently violated in many fields of natural science—big or small. Since "big science" research is carried out in groups and is frequently conditioned by the available facilities, true independence is hard to come by. At the same time, the graduate student may indeed make major and independent contributions to the instrumental part of his or her activity. However, academic departments are loath to recognize instrumental contributions as a significant component of a thesis, even if they are intellectually highly challenging. This should be changed.

- *Absenteeism.* The participation by the research physicist in experiments at SLAC is frequently controlled by the vagaries of the machine, the pattern of governmental financial support, and other factors not under laboratory control. Thus, a faculty member and his or her students may need to absent themselves from their teaching duties, with burdens consequently being placed on their colleagues. In most user institutions, this matter requires multiple teaching loads when the researchers return to campus. Universities will have to decide whether a price they are willing to pay for participation in big science should be a greater flexibility in assigning teaching duties.

- *Quality of Local Facilities.* It is clearly desirable, from the academic point of view, to minimize the absences of faculty and students from campus while they are collaborating in big science off campus. This would maximize the interaction of these individuals with their fellow academic colleagues and their contributions to academic life at the home institution. However, to make this practical, the home institution cannot be permitted to let its supporting facilities—shops, computing facilities, technical staff assistance—deteriorate; if that happens, the academic people have no choice but to carry out an even larger part of their work at the "big science" laboratory. Recently, the quality of support facilities at universities in the United States has slipped unacceptably. The blame for this must be divided between the sources of government support and the academics themselves. Financial support by the government has not grown as rapidly as most people had hoped it would, but, at the same time, the principal investigators at most academic institutions have been less than responsible in using available funds for augmenting the academic personnel, rather than building up the supporting structure.

There has been a tension between big science and the universities when it comes to the question of whether there should be a university or a branch of a university directly located at a big science facility. This is, of course, not an issue at SLAC, since Stanford University is in the fortunate position of being able to accommodate a 2-mile accelerator on its 17-square-mile campus. It should be noted, however, that when the University of Chicago attempted to establish a branch campus on the site of the Argonne National Laboratory, a great howl arose from many of the Midwestern universities, who claimed that the University of Chicago was trying to take unfair advantage of the Argonne resources relative to its sister academic institutions in the Midwest. At the same time, there is criticism—some of it merited—that the intellectual atmosphere at some big science facilities could be greatly improved if the relations to universities were more intimate. Attempts have been made in this direction, for instance, with respect to the branch of the State University of New York at Stony Brook, just outside the gates of Brookhaven National Laboratory.

The Future

The preceding discussion illustrates some of the social problems of big science, but these are not clearly fundamental, and they can be solved if physicists and administrators are willing to be flexible and accommodating in solving problems brought on by the requirements of advances in the science or the technology. One should note that, despite this dramatic evolution, the interval between major discoveries in elementary-particle science has not changed very much. If one makes a substantive list of such events for this century, one can point out perhaps one or two dramatic discoveries per decade that have truly changed our view of the inanimate world.

Thus, despite all the required changes and innovations during the past decades, progress in big science has been good, and, in particular, the motivation of practitioners of basic big science has remained the same as that of physicists many decades ago: to investigate nature in smaller and smaller dimensions and to build the tools necessary to do so. Will this situation persist to the next range of machines, which will probably require investments in the billions of dollars? I hope very much that it will, but there are some danger signs. When it was decided, within the various scientific councils in the United States, to proceed to the SSC as the next step in particle colliders, most people felt that the social environment would not be significantly different from that pertaining to the present laboratories, such as Fermilab or CERN.

In fact, the investment in the SSC project is larger by perhaps a factor of three than that at Fermilab. The annual expenditures for high-energy physics during SSC construction and thereafter would only approach in real cost those already experienced in the 1960s and 1970s. Yet hardly anyone foresaw how in the political climate of today, the jump in investment for a single instrument into the multibillion-dollar range would induce new social and political factors. The site-selection process for the SSC has brought out a fierce political competition that is based on what I consider to be a highly exaggerated perception of the promise of the SSC for the local economy. The spokesmen for the SSC have accentuated this misunderstanding by overemphasizing its potential for industrial spin-offs or contributions to international "competitiveness."

The price tag for the SSC is sufficiently large that there are many industrial concerns interested in moving the management of that machine out of the traditional academic, nonprofit pattern. In the United States, there has been alleged favoritism by the government in giving contracts without competitive bidding, particularly by U.S. Postal Service and the Department of Defense. This public outcry has been translated into pressure for "competition" in the management of the SSC, and such pressures have distorted the traditional pattern of the support of science. New science endeavors, for big or small science, spring from the desire of scientists to advance knowledge; this results in efforts to obtain support for those endeavors, and government will either support or not support such initiatives by the interested scientific groups. Now, in the name of competition, the situation has been reversed. Attempts are being made to mold the establishment of new scientific facilities into the norm applied to acquisitions or purchases: a "need" for such a new facility is established in the abstract, and then the illusion is created that such a new facility can best be built by those who make the most favorable response to a request for a proposal. Whether this trend (which is based on a profound misunderstanding of the nature of the scientific endeavor) can be reversed, considering the high price tag of big science, remains to be seen.

Another unforeseen result of the high cost of big science is the increasing competitiveness among scientists in different disciplines, which I mentioned earlier. Governments cannot expect that practitioners in different scientific disciplines will be equally enthusiastic about new initiatives by any one science, and decisions about the future of big science initiatives therefore cannot be made by a consensus of all scientists, let alone the public. There is no objective way to assess "values in science," notwithstanding the serious efforts made by some to quantify "criteria of scientific choice." Such intangible values as intrinsic and basic scientific interest simply cannot be compared with such predictable values as the expected economic benefits to society. Thus, the high political visibility and the public clamor surrounding the establishment of multibillion-dollar scientific facilities have cast a shadow over the future of very large instruments for basic science. I hope very much that this shadow can be dispersed. I see little valid economic reason for its existence, nor do I see any reason why the basic scientific spirit of inquiry cannot be maintained, notwithstanding the methodological changes that are required in view of the complex technology of modern big science.

Critics of "big science" often maintain that it is driven by a quest for bigness for the sake of bigness, and that more thought and analysis could obviate the need for the next generation of instruments. I disagree. Scientific knowledge must have an experimental or observational basis; nature is not limited to all that is reasonable and internally consistent. If the evolution of high-energy colliders is terminated by the social or political difficulties or by some perceived financial limit, then one of the most successful creative human endeavors of this century will have ended.

Technical Limits for High-Energy Proton and Electron Colliders

Werner Heisenberg's uncertainty principle requires that particles of very high energy be used to explore the nature of matter in its smallest dimensions. Thus, inexorably, progress in our knowledge of the microstructure of matter is linked to the technology of producing and using particles of higher and higher particle energies. Moreover, we now know that an understanding of the origin of the cosmos is linked to an understanding of physics at the smallest distances; the processes responsible for the generation of matter at the earliest time in the evolution of the universe take place at those high energies that are now brought under investigation with high-energy accelerators.

I hasten to add that elementary-particle physicists who wish to study nature at the smallest distances are not entirely dependent on high-energy physics accelerators and colliders. There are other complementary and competitive approaches. First, there are experiments using cosmic radiation as a source of particles. These constitute important tools not only for gaining an understanding of cosmic processes but also as sources of particles of higher energy than appears within conceivable reach with accelerators. Then there are experiments deep underground to study interactions of neutrinos that penetrate the earth's crust or to search for the existence of extremely rare decay processes of such particles as the pro-

A talk given as the Werner Heisenberg lecture given in Munich, Germany, March 13, 1991.

ton, which heretofore had been believed to be stable. Then there are experiments in which extremely high *precision* can be substituted for very high interaction *energy*. The reason for the power of such experiments is the existence of "virtual" processes, which can have demonstrable but very small effects at very low energy. Again, we have learned from Heisenberg that nature can borrow energy for a very short time; thus, the existence of high-energy states can contribute small effects to the reaction rates of low-energy processes. The success of this approach is critically dependent on precise theoretical understanding of the processes that are being investigated. If we can make very precise predictions about the outcome of specific experiments, then small deviations from such predictions can be associated with the existence of higher energy states. At this time, the only theory that meets the requirement of being understood with such extremely high precision is quantum electrodynamics—that is, the theory that describes electricity and magnetism under quantum-mechanical principles. Here, agreement between theory and experiment has been as exact as the precision of measurements permits, over a range in scale from cosmic distances down to dimensions on the order of 10^{-16} centimeter. This agreement notwithstanding, the search for small deviations in electrodynamic processes continues to be worthy of investigation.

While nonaccelerator approaches to physics at the smallest distances have indeed been—and continue to be—important, it is fair to say that their recent impact on increased knowledge of elementary particle physics has been small, compared with that produced by accelerators and colliders. The long-looked-for decay of the proton has thus far been elusive, but we conclude that, if the proton is unstable, its lifetime is well beyond 10^{33} years. This means that detectors operated underground must contain at least 10^{35} protons and therefore weigh at least 150,000 tons, if, even assuming efficiency of 100 percent, 100 such decays are to be observed per year. Selected cosmic-ray events—in particular, those associated with the recent supernova explosion—have generated identified bursts that have narrowed the limits on the value of neutrino masses. Neutrinos from the sun may also provide information on such masses.

The reason why cosmic radiation is limited in its utility for high-energy particle physics is the relatively low flux of incident particles as the energy of the particles is increased. For instance, the cosmic-ray flux of particles corresponding to a collision energy to be produced by the Superconducting Supercollider south of Dallas, Texas, or the proposed LHC at

CERN in Geneva, corresponds to roughly 1 particle per square kilometer per year—not a very useful flux for experimenters.

The totality of all the experimental information gathered with the help of accelerators and colliders, from observations on the cosmos, and from precise analysis of lower-energy experiments, is consistent with what has now been designated the "standard model." This model describes nature to be constituted of two families of particles that, down to a distance of 10^{-16} centimeter, are still pointlike and fundamental; these are designated *leptons* and *quarks*. In turn, the interaction among these particles is quantitatively described by the model, and the carriers of such interactions are themselves particles of reasonably well-established characteristics. These carriers are called *gluons,* for interaction among quarks; intermediate *vector bosons* carry the weak interaction, affecting both leptons and quarks similar to the *photon*, which mediate electromagnetism.

Yet the standard model *cannot* be complete, or even precisely accurate. There are reasons for this conclusion, both aesthetical and physical. Aesthetically, the problem is first that gravity, whose influence is observable only for large-scale objects, has not been incorporated in the standard model, and the theoretical conjectures on how to incorporate it are wildly beyond the current reach of experimental verification. Further, the standard model contains more than 20 arbitrary numbers that have been incorporated in the theory but whose physical origin is not as yet understood. Yet, at some level, the value of these parameters must reflect some heretofore unknown physical phenomena. Examples of such parameters are the masses of what are still believed to be the fundamental constituents—that is, the leptons and quarks, and the masses of the carriers of the interaction between them. Then there are the so-called coupling constants, which designate the strength of the various types of interactions; until such time as physicists have been fully successful in "unifying" these forces, the strength of each of them is still described by a separate number.

This large number of independent numerical inputs into the theory violates the sense of aesthetics of the physicists. Historically, particle physics has been a search for simplicity. The thrust is to unify the widest possible classes of phenomena into a single overarching description. Our history, in this respect, is illuminating. As our search for the fundamental building-blocks of nature proceeds, some overall simplifying notions are revealed. However, just as soon as some success in uncovering unifying assumptions has been gained, then new discoveries have proliferated new

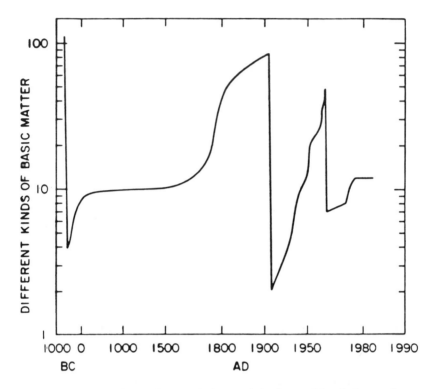

FIGURE 1. *People have always tried to explain the world as being made up of a limited number of different basic kinds of matter. Until a thousand years ago, the basic types of matter were considered to be earth, air, fire, and water. By about 1900, the basic types of matter were thought to be the almost 100 different chemical elements. At present, we believe there are about a dozen types of basic matter—namely, the leptons and the quarks.*

entities. Another new theoretical "breakthrough" is then needed to reestablish understanding in terms of a smaller number or parameters. Figure 1 illustrates this trend in history. It is an *absolute certainty* that, at energies now within technical reach of accelerators, our understanding will make it possible to decrease the number of independent parameters in the standard model. A vehicle must be discovered from which the diversity in masses of at least some of the constituents within the standard model can be derived.

For all the reasons discussed, there is an urgent desire for particle physicists to extend their reach to higher and higher energies, and devices

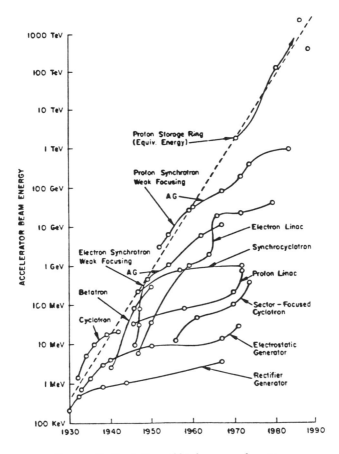

FIGURE 2. *Evolution of high-energy beams.*

that bring high-energy beams into collision will remain the primary tools to satisfy that quest. Figure 2 indicates the evolution of the energy of such devices in history, and we still see a pure exponential growth—about one decade in energy for each ten-year interval. However, this exponential growth has been attained not by simply scaling up devices that use well-established principles but, rather, by a succession of technologies that were reduced to practice once an earlier technology had been exploited to its limits. The question now facing us is to explore whether, and for how long, this exponential growth can continue, and what the limits are that inhibit that growth. No exponential growth—be it population increase on earth or the energy of accelerators—can endure forever.

We can talk about limits of growth in four classes:

1. fundamental limits;
2. economic limits;
3. limits set by the required data rates that can support practical experiments; and
4. limits set by the large "background" of old phenomena over new processes.

Fundamental limits imply that laws of nature prevent attainment of new parameters. *Economic* limits arise when the estimated costs of new projects appear to be too extravagant for society to support. *Data rate* limits arise when the information flow is too slow for the human mind to draw conclusions from the data in a reasonable time. *Background* limits arise when, with increasing energy, the yield of new phenomena decreases sharply, while the yield of well-known reactions does not.

Let me postpone discussion of the fundamental limits to the main part of this lecture. Let me say here only that the biggest change in the use of accelerators, which has maintained the exponential growth shown in Figure 2, has been the shift from accelerator techniques in which beams impact stationary targets to the use of colliding beams.

Let me explain. The reach of an accelerator into the unknown, and specifically its ability to discover new objects of increasing masses, is not defined by the *energy of the incident beam* but is defined by the *energy available in the collisions among the most fundamental constituents*. These two quantities can be different for two reasons illustrated in Figure 3: one is that, if a particle at high energy strikes another particle at rest, then some of the energy is used to move the center of mass of the combined system ahead and another fraction is available for the energy of the collision. It can be shown that, at the high energies that are now of interest, the energy available in the collision is proportional only to the square root of the incident energy, and thus, as we go to higher and higher energies, the fraction of energy of interest to the physicist relative to the incident beam energy diminishes correspondingly. The second reason is that we now know that protons and neutrons are not fundamental particles but are composed of quarks and gluons. Thus, if protons are in collision, then the energy in the total collision has to be shared among these more

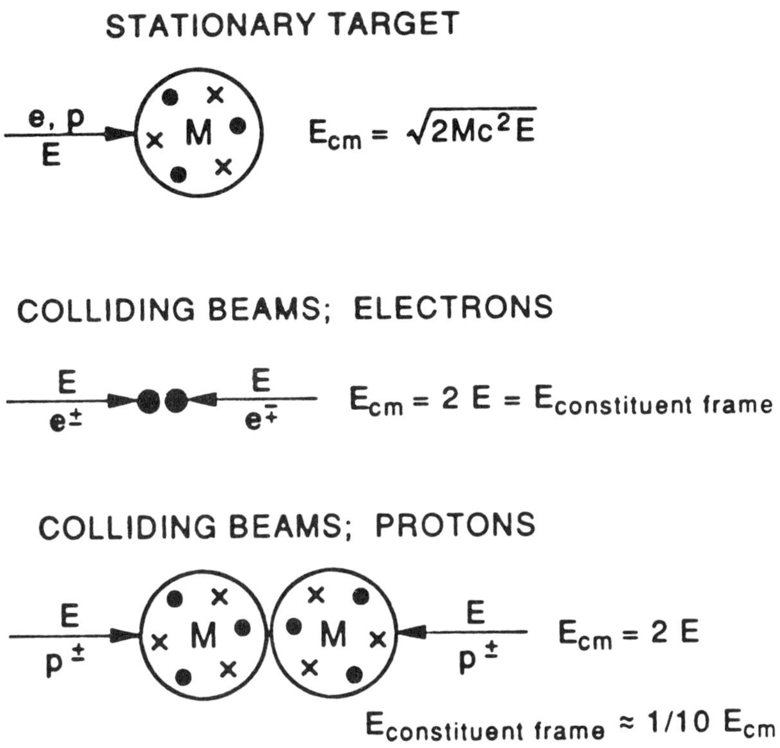

STATIONARY TARGET

$$E_{cm} = \sqrt{2Mc^2E}$$

COLLIDING BEAMS; ELECTRONS

$$E_{cm} = 2E = E_{constituent\ frame}$$

COLLIDING BEAMS; PROTONS

$$E_{cm} = 2E$$

$$E_{constituent\ frame} \approx 1/10\ E_{cm}$$

FIGURE 3. *Colliding beams versus beams on stationary targets, illustrated for electrons and protons. The center-of-mass energy (E_{cm}) as well as the energy in the frame of the fundamental constituents (quarks, gluons, and electrons) are illustrated.*

fundamental constituents, and therefore, the energy available for each collision among fundamental particles is decreased. Since electrons, being one of the family of leptons, are still believed to be fundamental at the highest collision energies that now appear to be accessible, this energy-division factor does not apply to electron colliders, but it does apply to colliders using protons or other nuclear particles.

The first factor can be made to disappear if we use colliding-beam machines rather than having beams impact on stationary targets; in that case, the center of gravity of the system in collision remains at rest, and the total beam energy is available for the collision. Thus, colliders for high-energy particle physics that have been designed recently have all been of the colliding-beam type, some for electrons and positrons, some for collisions between electrons and protons, and some for protons or antiprotons. For the latter, the second derating factor applies, but it does not if only electrons and positrons are used. I can therefore replot Figure 2 in terms of the energy in the "constituent frame," where by that I mean the energy available in collisions among the most fundamental constituents; the result is shown in Figure 4.

As far as economical limits are concerned, I note that, in parallel with the exponential growth shown in the previous figures, there has been a decrease in cost per unit of energy that is almost as dramatic as the growth in energy. Specifically, as the laboratory beam energy has increased by seven orders of magnitude since 1930, the cost per unit of energy has decreased by more than five orders of magnitude. The result is that the cost of a given installation has increased by perhaps a factor between 10 and 100. That increase has fueled a debate in the industrialized countries about whether society should spend sums as large as, for instance, the projected cost of the SSC in the United States—beyond $8 billion—for an instrument of basic science. One must, however, agree that the direct cost of basic science in our economic life is still sufficiently small that economic limits are a matter of policy rather than economic necessity. Costs of basic science are still a small fraction of 1 percent of the national product of the nations concerned. I would like to remind you of the joking suggestion, made in the early 1950s by Enrico Fermi, that an accelerator should be built in the vacuum of outer space as a complete ring encircling the earth. In 1956, I had a discussion in the Soviet Union with Artem Artzimovich, the great Russian plasma physicist. At a lag in the conversation, Professor Artzimovich asked me whether I had made a cost estimate of Fermi's proposed machine. I made a quick calculation and answered, "The sum of the U.S. and Soviet military budgets would pay for it in two years." Professor Artzimovich changed the subject.

Now let me turn to data-rate limits. There is a fundamental law of physics called "unitarity," which demands that the cross section of producing a new entity of a given mass decreases inversely as the square of that mass. Therefore, if the data rate of producing such a new object shall be maintained at a reasonable value, then the required so-called luminosity, which is the event rate per unit of cross section, must increase with the square of the energy. Technically, increase in luminosity depends on the increase in the intensity of the beams in question and also on the den-

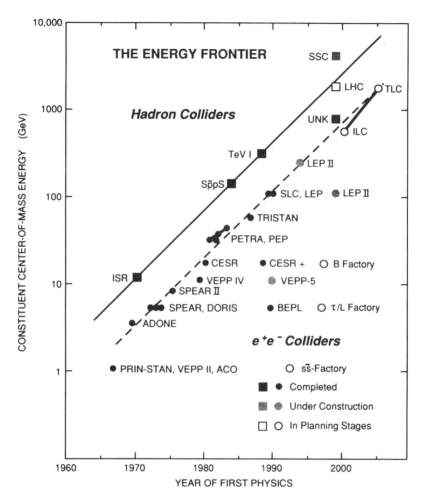

FIGURE 4. *Colliding-beam machines on the energy frontier of particle physics.*

sity of the target that is hit by the beam. In the case of colliding-beam machines, this is the density of the opposing beam. When, for the reasons indicated earlier, machines in which beams strike stationary targets were replaced with colliding-beam devices, the price paid was a dramatic decrease in the density of the target—ordinary matter has densities of perhaps 10^{22} atoms per cubic centimeter, while traditional particle beams in

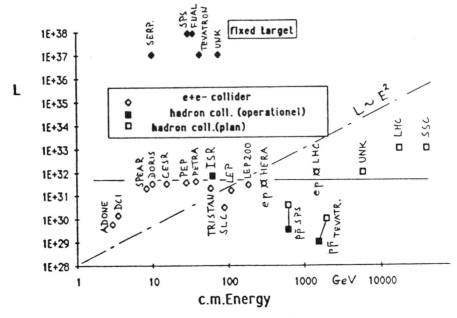

FIGURE 5. *Luminosity versus center-of-mass energy.*

accelerators have densities of perhaps 10^{13} particles per cubic centimeter. As the required energies increase, this is no longer tolerable, and therefore much of the engineering thrust of designing modern colliders has been to increase the density of the two interacting beams. Figure 5 indicates the luminosities that have in fact been reached in recent colliders; it is seen that the goal of having the luminosity increase at the square of the energy has not been reached, indicating that this is not an easy undertaking.

Then we have the problem of backgrounds. Here, the situation is drastically different between colliders employing electrons and those employing protons. For protons, the problem is that the total cross section with which proton beams interact with one another continues to increase slightly with energy, while, as noted before, the cross section of the interesting processes decreases with the square of the energy. This is illustrated in Figure 6. The reason for this is basic: since the proton is a composite particle composed of quarks and gluons, most of the cross section involves corrective interactions among their constituents. Such collisions

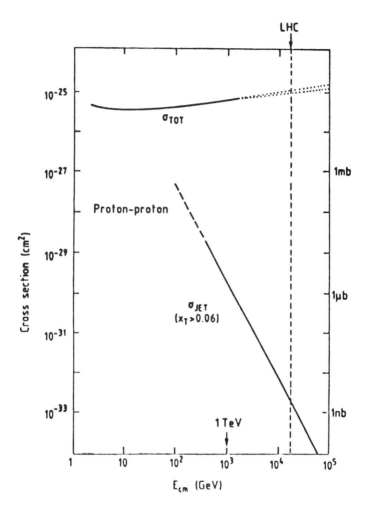

FIGURE 6. *Proton–proton cross section as a function of the center-of-mass energy. The dotted lines represent the extrapolations of the total cross section. The continuous line is the cross section for producing a jet having transverse momentum $p_r + x_l\,E_{cm}/2 > 0.03 \times E_{cm}$ [10].*

are known as "soft" collisions. On the other hand, those interactions that produce new objects of large mass continue to decrease with the square of the energy. Thus, the ratio of interesting signal-to-background for proton colliders becomes an ever-increasing problem. To cope with this problem, it is necessary to design complex detectors that surround the in-

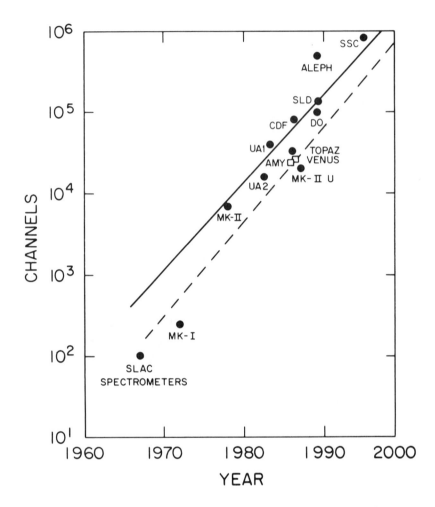

FIGURE 7. *Evolution of the number of data channels in detectors of reaction products of collisions produced by high energy colliders.*

teraction point as completely as possible in order to register all interesting events, but, at the same time, such detectors have to be segmented into sufficiently small subunits so each individual particle can be tracked and sorted out. This, in turn, produces a dramatic increase in the total data-handling problem associated with modern proton colliders. Figure 7 indicates the total number of channels that have recently been required for typical detectors.

Let me illustrate these numbers as they pertain to the American SSC. The colliding beams of that device are expected to generate about 10^8 interactions per second, with the bunches colliding every 10 nanoseconds. Each collision generated several hundred secondary particles, which generally have to be tracked and identified. Thus, each event, when duly recorded, might require the registry of perhaps 1 megabit of information. Yet the luminosity is expected to yield fundamentally new events corresponding to the formation of masses in the range of several tera-electron-volts (TeV) at a rate of perhaps 100 or 1000 per year. Thus, the number of "interesting" events is perhaps 1 in 10^{12} of the total hadronic reactions. In addition, the recording of a single event requires the writing of perhaps 10^6 or 10^7 bits of information. Thus, if everything the SSC produces is to be recorded, this would imply writing 10^{15} bits per second. This is unattainable by foreseeable technology, and, in fact, recording events at a rate something like a few per second is much more reasonable. These raw numbers, of course, exaggerate the actual situation, since many of the background processes occur at such small angles that they escape down the beam pipes and are not observable at any rate. Yet this situation implies that recording must be "triggered" in such a way that all but perhaps one in 10^8 events are rejected. The experimental consequences of this circumstance have been studied extensively for processes that have been theoretically conjectured in the new energy range accessible to the SSC, and detector designs, and the triggering algorithms that go with them to isolate such processes, have been conceived. Although, when looking at a specific process, the analysis problem appears to be tractable, the tantalizing question remains whether and how one can intelligently develop selection criteria that will not also prohibit the discovery of phenomena that have not been conjectured in advance to occur. In other words, what is the extent to which we are negating the discovery potential of very-high-energy proton machines by the necessity of rejecting, a priori, the events we cannot afford to record?

The analysis situation is vastly different for electrons. Here, the cross section for both the reaction of broad interest as well as those events following traditional expectation are both expected to decrease together, as shown in Figure 8. Thus, there should be much better clarity in distinguishing new events from background using electron–positron colliders.

Physicists have carried out extensive calculations that simulate the process of discovering new generations of particles. These simulations take into account the backgrounds produced from known processes, and

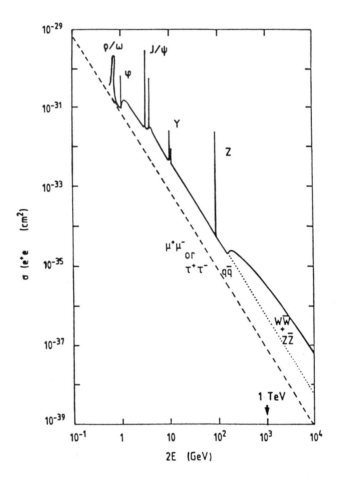

FIGURE 8. *The cross sections expected in electron-positron collisions as a function of energy.*

they include such detailed characteristics as efficiency, angular accep-
tance, and resolution of a particle detector. From such calculations, a
"range of discovery" can be defined over which each new collider could
have power to observe new objects of certain characteristics in a given
range of mass. Not surprisingly the SSC, being the highest-energy collider now technically feasible, has the largest range of discovery. However, there are holes in its coverage where well-established processes

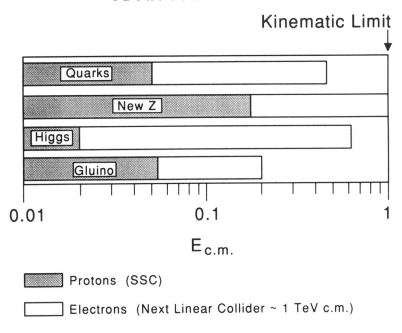

FIGURE 9. *Discovery reach of proton and electron colliders.*

simply cannot be distinguished from new phenomena. The technology of electron–positron colliders that is now available limits the expectancy for discovering masses as heavy as those detectable by the SSC, but their range of discovery, as far as it does reach, is more comprehensive and clean. Figure 8 compares the "range of discovery" of the SSC with that of a conjectured electron–positron collider of maximum energy near 1 TeV. In Figure 9, that range is plotted for the discovery of various new particles relative to the "kinematic limit" for each machine, meaning the limit set by the energy available for production of the new particle. Obviously, the electron machines look better on a *relative* basis; however, on an *absolute* scale, the reach of discovery of the SSC is far greater. How, then, can we extend the energy range accessible to electron–positron colliders?

Extension of the energy of electron–positron colliders requires fundamental innovation. The situation has been characterized facetiously by Carlo Rubbia, Director General of CERN, by saying, "We have to choose between a collider we do not know how to build [meaning electrons] and one we do not know how to use [meaning protons]." The situation may not be as dramatic as that but, from the fundamental point of view, the technological energy limit for proton colliders indeed appears to be set currently by detector technology, while the technology for building an electron–positron collider for higher energies is not in hand.

What are the technical factors? Both electrons and protons, when confined to nearly circular orbits, are expected to emit high-energy x rays called synchrotron radiation, and the emission of such radiation impacts on the engineering. The rate of radiation under otherwise equal conditions is smaller by a factor of almost 2000 for protons than for electrons, that number being the ratio of the mass of the proton to that of the electron. Although this effect is much smaller for proton machines, as far as the energy loss represented by that radiation is concerned, it is significant in that the radiant flux impinges on the cold superconducting magnets and vacuum chamber. Thus, although the radiating power is small, it might have to be dissipated at very low temperatures and thus require large increases in refrigeration power. Therefore, this problem may become severe at energies significantly above that of the SSC.

For electron–positron colliders, synchrotron radiation is already a dominant design factor for all machines now is use. Specifically, the energy loss per turn caused by synchrotron radiation varies with the fourth power of the electron energy divided by the radius of the orbit. This energy loss must be compensated for by large amounts of radio-frequency power supplied to the electrons. Thus, the dominant costs of an electron collider consist of one term proportional to the radius of the machine, and another term that is proportional to the fourth power of the energy divided by the radius. To optimize the design of such a collider, these two terms should be matched by having the machine radius increase proportional to the square of the energy. It is for this reason that the electron collider LEP at CERN is limited to about 200 GeV, while the LHC for protons, if it is to be built, could reach an energy about 50 times greater. Both the cost and the size of an electron–positron *storage ring*, therefore, grow with the square of the energy; therefore, one can conclude that LEP is probably the last machine in the world that will be constructed on that principle.

The electron–positron colliders now appearing capable of reaching higher energy are "linear colliders." Here, beams accelerated in two linear accelerators are brought into collision. Such machines do not exhibit

FIGURE 10. *SLAC Linear Accelerator Site. Note SLAC Linear Collider (SLC) shown as a dotted line.*

the limits of storage rings, as set by radiation just discussed, but they invoke other fundamental limitations.

In a storage ring, the particle bunches are literally "recycled," leading to bunch collision rates in the megahertz range. In a linear collider, they are discarded after each passage. This fact, combined with the need to increase the luminosity of such machines quadratically with energy, leads to the requirement to increase the beam density dramatically as the energy is increased.

The path to increased electron-beam densities has been initiated in the SLAC Linear Collider (SLC). An aerial view is shown in Figure 10. Here, electrons and positrons are produced, accelerated to 50 GeV, focused, and brought into collision at densities approximating 10^{19} electrons

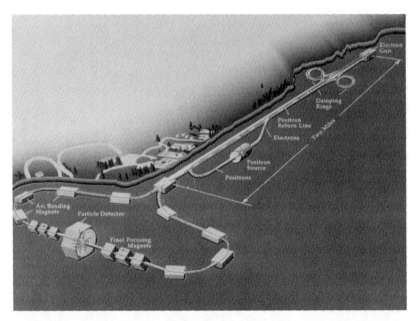

FIGURE 11. *The principal components of the SLAC Linear Collider.*

per cubic centimeter. The principle of that machine is not scalable to higher energy, since it takes electron and positron beams accelerated in a single machine and bends their orbits into collision, as is shown in Figure 11. Once the electron energy substantially exceeds that of the SLC, then the radiation produced in the bending magnets would cause excessively disturbing effects on the beams. Thus, linear electron–positron colliders of the future that go beyond the energies attainable at the SLC and LEP must employ an "honest" linear-collider principle, in which two independent linear accelerators hurl particles at one another. Figure 12 compares the principle of the SLC now operating at SLAC with that of an "honest" linear collider. The question is, Are there any evident limits to that "honest" approach? Reluctantly, I would deem the answer to that question to be "yes."

There are, of course, economic bounds. If one simply scales the practice of existing linear accelerators to energies into the TeV region, then costs would exceed those of the SSC. Thus, technologies yielding econ-

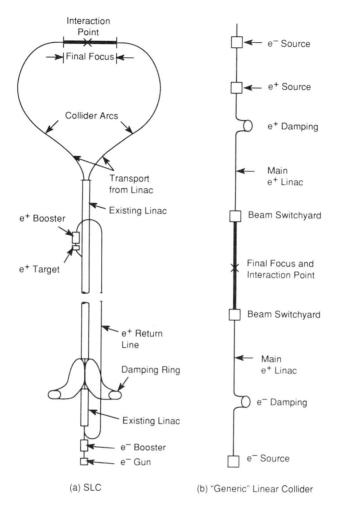

FIGURE 12. *Building blocks for e$^\pm$ linear colliders.*

omy superior to that of existing linear accelerators would have to be developed. Fortunately, there are several promising paths in sight by which this might be accomplished. Beyond simple fiscal economy, there is the matter of length. It is unattractive, if not prohibitive, to build a pair of linear accelerators to attain energies of many TeV—that is, many thousands of GeV of electrons and positrons—by simply scaling the length of the present 50 GeV linear accelerator at SLAC, which is already 3 kilometers long. Thus, a higher energy gradient must be attained beyond that of 17 MeV per meter, which is now obtained with the SLC at SLAC. Again, this appears possible; in fact, a gradient of 200 MeV per meter is not impossible for conventional structures, in which high radio-frequency powers are guided in a disk-loaded waveguide to produce accelerating electric fields.

However, the maximum energy gradient is not necessarily the most economical choice. The reason is that currents in the wall of the accelerating structure produce radio-frequency wall power losses that scale linearly with the energy gradient, while the cost of the accelerating structures and tunnels scales linearly with length and, therefore, varies inversely with that gradient. Thus, minimum cost requires a compromise between low and high accelerating strength.

In view of the above, attaining the requisite *energy* in an electron–positron linear collider is limited by economics rather than by technology. This, however, is not the case with respect to the *luminosity* or the *data rate*. As noted earlier, the luminosity of colliders of the storage-ring type has not kept up with the requirement of increasing with the square of the reaction energies.

Although the limits to the luminosity attainable by linear colliders are of a different physical nature than those for storage rings, there are limits nonetheless. In essence, the luminosity is proportional to the product of the number of particles in each bunch that are undergoing collisions, divided by the cross-sectional area into which these particles can be compressed, multiplied by the frequency with which the bunches undergo collisions.

There are two basic limits that define how many particles can be crowded into a given cross-sectional area. First, there is a limit analogous to a well-known limit on the brightness of optical images. According to the second law of thermodynamics, the brightness of an image produced by a light beam can never be brighter than the source from which the light beam originated. By brightness, we essentially mean the intensity emitted or received per unit of area. In optics, this means that, when we focus light from the sun on to a small spot, the image can never become hotter than the surface of the sun. For accelerators, it means that the number or electrons emitted per unit area of the original source cannot be focused into too small an area. If the beam is accelerated, then the final spot can be decreased in area by an amount related to the acceleration, but it is limited nevertheless.

In practice, it is difficult to produce a focus that is as good in quality as the fundamental limit implies. There are many factors at work that dilute the beam quality; most of these are related to the precision to which the accelerating structure, which must be many miles in length, can be manufactured and aligned. Analysis indicates that we are talking about micrometer, or even submicrometer, tolerances. Moreover, these tolerances become more stringent with increasing energy of the beam, varying with a large power of that energy.

The second problem has to do with beam–beam interaction at the final focus, where the bunches of electrons and positrons collide. The problem is that, as the beams are focused to the extremely small diameters required, the magnetic fields at the edge of the beam become very large, literally many millions of times larger than the earth's magnetic field. When exposed to such an intense field, the electrons and positrons will radiate electromagnetic energy, thereby spreading out the energy remaining. Moreover, in such intense electromagnetic fields, the electrons will also produce additional pairs of charged negative and positive electrons, which would seriously increase the presence of confusing background events in the detectors.

Thus, there are limits to the number or electrons and positrons that can be concentrated into a single bunch without producing adverse effects. The only variable left over to increase the luminosity is the frequency of collisions between the bunches.   There are two ways to increase the frequency of collisions: the first is the repetition rate at which we feed pulsed radio-frequency power to the accelerating structure, and the second is to increase the number or particle bunches accelerated within each radio-frequency pulse. The first approach faces the limit set by the electric-power requirements of the collider. If we increase the frequency by which we must feed energy into the accelerating structure, then the power requirement goes up. If fact, analysis indicates that, with one particle bunch per radio-frequency pulse, there is a direct relationship between the power that must be supplied to the beam, the required luminosity, and the tolerable energy-broadening of the beam due to the beam–beam interaction.

If several particles bunches are accelerated per radio-frequency pulse, then limits are produced by instabilities generated by radial deflecting fields produced by a leading bunch, which disrupts the following group of particles. Such instabilities can be limited by ingenious design of the accelerating structure, but they cannot be eliminated. All these relationships are very complex, but they are persuasive that the next generation electron–positron linear colliders, beyond the SLC at SLAC, cannot conservatively aspire to an energy much beyond perhaps 300 GeV per beam; building a linear collider intending to match the discovery reach of the SSC would require beam energies well above 1 TeV per beam—and we do not know as yet how to do that.

There remains the possibility of making the radio-frequency accelerating structure superconducting. At radio frequencies, superconductivity does not imply zero surface resistance but, rather, losses that decrease sharply as the temperature is lowered to that of liquid helium. If superconducting structures are employed, then pulse length can be very long, and thus many electron bunches per pulse are feasible. However, there is a technical limit to employing such structures that for the time being, appears insurmountable. This is the combination of the relatively low energy gradient attainable with the use of such structures together with the high unit cost of their manufacture. The practical limit in gradient for superconducting structures is, at present, 20–30 MeV per meter, and, theoretically, that limit cannot be exceeded significantly without surpassing the critical magnetic field at the surface of the structure, above which superconductivity cannot be maintained. Thus, TeV-range accelerators would approach a length of 100 kilometers. This length, combined with the current cost of producing such structures with their cryogenic envelopes—about $100,000 per meter—means that superconducting linear colliders remain an object for the more distant future.

When elaborated quantitatively, these considerations (which I have outlined here only qualitatively) show that, by using known technology for electron–positron colliders, we may be able to reach the range of collision energies around 500 GeV, but we cannot go much beyond that without some drastic changes in basic assumptions. Quite separate from these considerations, it is necessary to develop economical power sources that can feed such accelerators; general scaling relationships indicate that such sources should be microwave devices feeding power in the wavelength range of a few centimeters.

These remarks have identified the existence of possible severe limits to the expansion of colliders, both for electrons and protons, into new realms of physics. These limits may simply be symptomatic of a paucity of ideas and a lack of imagination, and one can hope that, as has been the case before (and as illustrated in the growth curve in Figure 2), new technologies will invalidate such fears.

Many suggestions have been promulgated to increase particle energies and energy gradients, based on such advanced technologies as lasers and plasmas. However, progress demands not only an advance in energy but, at the same time, advances in luminosity and economy. Thus, we require not one but many simultaneous inventions. In fact, most such "high technology" suggestions for providing higher energies have serious deficiencies, in terms of beam quality and efficiency with which energy from the source of electric power can be transferred to the beam.

If it were to be true that advances of colliders much beyond the reach of the SSC illustrated in Figure 9 were not possible, then one of the most exciting and creative epochs in scientific endeavor would have come to an end. Since 1930, equivalent accelerator energies have grown by seven orders of magnitude, and the information gained has illustrated the physical basis of events ranging over billions of years, from the first fractional seconds of the big bang to phenomena of great relevance today, including the understanding of nuclear energy. In other words, the advances in the collider arts have generated knowledge of the basic processes in the history of the universe in less than a single century, a compression of time by a factor of perhaps 100 million. Let us hope that our ingenuity can extend this creative epoch and allow us to overcome the limits we have discussed.

Science, Technology, and the Arms Build-Up

The arms race, in particular the competition in nuclear weapons between the Soviet Union and the United States, threatens the very existence of civilization. Yet one must admit, albeit reluctantly, that the nuclear balance between these two powers deserves at least partial credit for the absence of all-out hostilities for the longest period in recent history. But, even in the absence of total conflict, the increase of potential devastation and the growth in the number of states that possess nuclear weapons have created tensions that increasingly overshadow all other human concerns. Many hold science, which made possible the unleashing of nuclear energy, responsible for this development. But others look to science for the tools to eliminate this threat to all humanity.

Accordingly, a discussion of the relationship of science and technology to the arms race is timely. I will here only touch upon a topic of at least equal importance—the relation of science and technology to arms control.

Science and Military Devices

The relation of science to war and preparation for war has been what might be called a love–hate relationship throughout recorded history. Let

A Colloquium on Science and Disarmament given at Institut Francais des Relations Internationales, Paris, France, January 16, 1981.

me remind you of the story of Archimedes: In the defense of Syracuse against the Romans, Archimedes reputedly devised numerous instruments of war that contributed materially to delaying Roman entry into that besieged city for many years. When the conquest of Syracuse finally took place, in 212 B.C., Archimedes was engaged in "pure" science, drawing circles in the sand in studying basic geometry. When a Roman soldier approached him, Archimedes was reported to have uttered these last words: "Don't disturb my circles." This story illustrates that scientists wish to be recognized by all, friend and foe alike, for their contributions to pure science, irrespective of whether they also play an important—or even decisive—role in providing armaments to their home country. A vexing conflict emerges between two moral responsibilities of scientists in relation to armaments: their duty to their country, on the one hand, and their responsibility to prevent the abuse of lethal weapons, on the other.

This conflict is rooted directly in the nature of the scientific process. New science and technology evolve through a long chain: Basic scientific discoveries eventually lead to the development of specific new devices, and finally to their testing, production, distribution, and use.

Each of these steps reflects different needs and opportunities for control by human institutions. Most scientists and nonscientists would agree that *basic* science, motivated solely by the desire for improved understanding of nature, should remain unfettered. I would maintain that, however fallible humans and their institutions may be, decisions on how to live harmoniously with nature and its products will be made more wisely with an understanding of the workings of nature than without. Yet, once basic new revelations are apparent, applying restraints to their potential fruits, lest they be abused for asocial ends, is difficult. In regard specifically to the problem of controlling military arms, one finds that controls are very difficult to apply to early phases of the development chain. As a practical matter, particularly if "verification" (that is, policing of an arms-control agreement) is to be taken into account, controls can be applied primarily to the *test* phase of development, on the one hand, and to final deployment actions, on the other.

It has become trite to say that the rate of progress in arms control has been so small that the rate of technological advance of military systems has outpaced the gains of arms control. Yet I would argue that, had technological and purely military factors been the only ones in determining the build-up of arms, arms-control efforts might well have stemmed, and possibly reversed, the build-up of nuclear weapons. In truth, it is largely political factors that have made it infeasible thus far for arms-control efforts to prevail over the arms build-up.

Science in Relation to Other Factors Driving the Arms Race

Evolution of the tools of war, as facilitated by advances in science and technology, has aggravated the action and reaction cycles between nations and has led the competition between systems of offensive and defensive technology through successive stages. Yet, while such cycles are clearly demonstrable, it appears that the nature of the arms competition is, in fact, *not* dominated by such direct causal factors. Examination of the history of the arms race appears to demonstrate that nonscientific factors have pre-empted what might be logically predicated on technical or even military grounds. For instance, if one observes the arms competition between the United States and the Soviet Union, one finds very few instances of demonstrable action–reaction processes. Apparently, the decisions by both sides to proceed with arms acquisitions are driven by other forces. What are these forces?

Let me describe a few of them:

1. It appears clear that, in all countries—capitalist and socialist alike—there are strong *institutional pressures and interests* for augmented armaments by the institutions charged with developing and producing, or with managing and using, such arms. This commonality of pressures for arms expansion within different social systems does not mean that the sources for military expansion are identical, but it makes it difficult indeed to attribute such growth to single factors, such as the profit motive, communist or imperialist expansionism, interservice rivalry, and so on. We are dealing here with institutional inertia, or with other manifestations of historical persistence, that are inherent in any highly organized human activity. Institutions always find it difficult to produce only a fixed and limited amount of any one commodity; the producer always rationalizes why more of what he can produce is needed.

2. A second driving force in the arms race is what I will call *mirror-imaging*. Rather than reacting to a potential opponent's new military system by implementing specific countermeasures, the reaction frequently is, "We must have it too." This reaction tends to be relatively uncritical with regard to the technical merit of the item in question. The public rationale given, for instance, among other factors, for the United States to develop an improved hard-silo killer for its missile arsenal is that the Soviets have

developed enough striking power to threaten pre-emptively the U.S. Minuteman silos; yet the public rationale for the U.S. "silo killer" missiles disclaims any intent for a "pre-emptive" or "first" strike. A "mirror-image" response aims to silence the implied threat that the Soviets would receive *political* advantage and bargaining leverage out of these weapons, unless the United States could field a matching threat. Conversely, one of the motives for some Soviet deployments appears to be to "mirror-image" the United States rather than to provide a countermeasure to U.S. moves. For example, the Soviets continue to upgrade the technological complexity of many of their military systems, following the U.S. lead. Yet, in so doing, they at times inherit the maintenance and reliability problems that have so frequently plagued the U.S. systems in the past.

Very new and frequently spectacular technology is particularly sensitive to this aspect of the arms race. For example, the militarily highly dubious development of particle-beam weapons is largely promoted by the argument that "the other side is doing it," rather than by valid military–technical arguments. The difficulty here is that the political prestige engendered by a perceived military–technical "breakthrough" is simply too valuable a commodity for decision-makers to forgo, even if the true military significance is minor or nonexistent.

3.   Possibly the most serious driving factor in the arms race is *the asymmetry of perception*, resulting in uncertainty and fear. Each side in an adversarial situation tends to feel insecure and threatened by the technical developments of the opponent, particularly if the technical details are poorly known and if the doctrine under which they are developed is ambiguous. For instance, the claim that the opponent is not only striving for but has, in fact, attained nuclear superiority rather than parity is being heard both from U.S. and from Soviet spokesmen in the current state of U.S.–Soviet relations. Such spokesmen may, in fact, be sincere, yet both cannot be correct. In fact most discussions on superiority tend to be based on simplistic numerology. Items that are difficult to quantify, such as the reliability of allies, asymmetries in geography, access to oceans and the length and number of boundaries shared with potential adversaries, make "superiority" an even less tractable quality. As a result, the question of "superiority" is one of perception of potential adversaries rather than objective reality.

As a result of asymmetry of information and interpretation, it is not unreasonable that both sides in the current arms build-up feel genuinely threatened. Political leaders at times react to these perceptions by "saber rattling"—that is, by proclaiming their military prowess "second to none" in order to reassure their constituencies. At other times, in particular when military appropriations or other allocations of resources are at stake, political leaders emphasize the "gaps" in the armor of their country—that is, the areas of perceived deficiencies and weakness. Thus, the perceived rather than the real military situation greatly affects decisions about the acquisition of military weapons.

4. The development and deployment of arms generally requires very long and frequently uncertain lead times—well over a decade elapses between development for most systems and their completion and final deployment. Thus, military needs should be based on a combination of judgment of the opponent's intentions, evaluation of his current status, and intelligence projections from that base. The validity of those projections and their accuracy are impaired not only by the basic uncertainty in foreseeing future developments and future intentions starting from known current conditions but also by the ignorance imposed by the secrecy that shrouds those conditions. Thus, it is not surprising that judgments about future defense needs can span a very wide range between best-case and worst-case projections. This gap is amplified by secrecy, and secrecy therefore becomes a major factor in driving the arms race. I challenge national leaders, in particular those of the Soviet Union, to examine critically whether extreme secrecy on technical developments is really in the best security interest of their country, quite apart from the damage such secrecy does to basic human values and, incidentally, to the feasibility of negotiated, verifiable arms control.

5. The argument that, for the most part, political and perceptual factors rather than technical ones have driven the arms race must, in fairness, be balanced by the existence of what has been designated as the *technological imperative*. While a vast number of technical developments are, on balance, beneficial, there may be a rational assessment that a certain new development, if generally adopted, might result in decreased security for all; the strategically destabilizing influence of high accuracy of ballistic missiles is a case in point. Yet prohibiting or impeding the evolution of such a "counterproductive" technology under current conditions is generally infeasible. A development of what J. Robert Oppenheimer called a "technically sweet" military technology is hard to stop, either unilaterally or through negotiated arms control. Unilaterally, the argument "What will we do if the other side gets it first?" tends to prevail. In an arms-control negotiation, the difficulty in policing the evolution of a new technology

in its infancy is usually a block to agreement, unless very low standards of verification are acceptable, or unless adequate openness of the societies leads automatically to public exposure. Thus, new technology tends to generate its own momentum, and unilateral or negotiated control is impeded.

6. There has been much speculation on the escalating role of *arms-control negotiations* on the weapons build-up. I will not discuss this subject here in detail. I would like to conclude, however, that the arms-control efforts during the last decade have, on balance, led to positive results. In particular, the SALT I agreement has increased world security, and so has the Treaty on the Non-Proliferation of Nuclear Weapons. Also, the various achievements in less central areas of arms control have certainly made net positive contributions. In addition, the very existence of arms-control discussions, even those not leading to positive results, has maintained channels of communication and has permitted exchanges of data that, in the absence of such talks, would hardly have been possible.

Despite this net positive assessment of arms-control efforts, one should also acknowledge the negative by-products of arms-control negotiations. Arms-control negotiations and extensive ratification discussions have given undue political importance to the detailed numbers of military systems—in particular, those employing nuclear weapons—that their actual technical performance does not deserve. No analyst could reliably predict a specific dependence of outcome of military conflict on the precise number of nuclear weapons now deployed. Yet, as we debate about who has gained or lost in an arms-control agreement, the detailed numbers gain undue significance.

It is true that arms-control agreements tend to leave "uncashed bargaining chips" on the table. When a key military system has played a politically important role in the bargaining process in arms-control negotiations, then it is politically very difficult to abandon the system after the negotiation, once it has not been possible to ban its deployment as part of the agreement reached. The 54 U.S. Titans are a case in point. Although the recent accidental explosion in a Titan silo suggests that it might be advantageous to decommission the U.S. Titan force, this is politically very difficult, since their numbers played a significant role during the SALT II negotiations and in forming the political perception of U.S. strengths. Similarly, it was a matter of considerable difficulty for the United States to discontinue the deployment of those ABM forces permitted under

SALT I, however limited their strategic importance may have been. The Soviets have not as yet succeeded in doing so, and they are in fact putting their ABM deployment around Moscow through a new round of modernization, notwithstanding the limited potential effectiveness of this system.

The domestic debates in the United States (and presumably also in the Soviet Union) surrounding the arms-control negotiating process also related directly to the military build-up. This matter has several diverse aspects: For instance, during the ratification process in the U.S. Senate, the price for a favorable action voiced by some senators was an escalation of defense expenditures. Similarly, it is maintained that an arms build-up is essential to make it possible to "bargain from strength"—from example, for many years, the Soviet Union showed no interest in nuclear arms control, until a very substantial nuclear arsenal was at hand.

After the Soviet Union deployed (and continues to deploy) their mobile and accurate SS-20 medium-range ballistic missiles targeted against Western Europe, the United States proposed to station, on the soil of members of NATO, a number of missile systems that can reach the Soviet Union. However, members of NATO were willing to accept this more advanced "Theatre Nuclear Force" only if the United States agreed to pursue negotiations with the USSR on limiting such forces.

These examples indicate how interwoven arms control and weapons build-ups have become: We are hearing "No arms control unless we first arm," but also "No new arms unless we move to new arms control." While such coupling may represent current political reality, it breeds cynicism among the nonnuclear nations that the efforts of the Soviet Union and the Western allies are genuine in reaching meaningful and incisive arms control. On technical grounds, there is no basis for such linkage: *A good arms-control agreement increases the security of all participants*, irrespective of build-up in areas not controlled by the agreement.

All of these factors, combined with political developments, have led both the Soviet Union and the United States to escalate their nuclear arsenals to the potential of worldwide destruction. I should like to add, however, that the *rate* of growth of military expenditures of less developed countries is actually higher than those of the two superpowers, and that the fraction of GNP dedicated by some smaller nations to military pursuits is higher than that of the United States, or even that of the Soviet Union.

The two superpowers have at times, driven by the nonscientific, nontechnical factors listed above, procured large and expensive systems that are only of marginal value to their security. The MIRV build-up by the United States and then by the Soviet Union is a case in point. Most would agree that the security of both countries would be greater if neither had deployed MIRVs. Originally, the United States decided to develop and deploy MIRVs primarily as a penetration aid against Soviet ABM deployments; yet it is well known that Soviet ABM systems evolved at a rather slow and irregular pace. Economic arguments (cost per reentry vehicle) also played a role. Even when it became evident that Soviet ABM systems were not a very effective countermeasure against U.S. missiles, U.S. MIRV deployment continued. The U.S. land-based MIRV missile build-up was then matched and exceeded by the Soviet Union, even though the need for MIRV to penetrate U.S. defenses was clearly absent. Whether the Soviet motive in building up MIRV forces was to "mirror image" the U.S. build-up, or was to threaten pre-emptive attack against U.S. missile silos, or was simply to lengthen the potential target list, we can only surmise. *There is no doubt that, had arms-control efforts by the Soviet Union and the United States led to more incisive results at earlier times, the security of both nations and of the international community would now be greater.*

Strategic Doctrine: Limited Conflict

There is another factor that has militated against unilateral or successfully negotiated steps to limit deployment of strategic weapons; this is the vacillation and ambiguity of stated doctrine by the United States and the lack of clarity in the available statements of Soviet doctrine.

All declarations of U.S. strategic doctrine give highest priority to *deterring* initiation of *nuclear* war by an opponent by demonstrating to the potential initiator of such war that he faces an "unacceptable" outcome. This highest-priority goal has been supplemented during various U.S. administrations by listing additional objectives. *Unfortunately, once strategic doctrine wanders beyond the limited goal of deterring nuclear war, it faces severe internal inconsistencies.* For instance, U.S. doctrine tends to include provisions aimed at limiting damage, or assuring that U.S. damage would be less than that suffered by the adversary, *if deterrence were to fail.* Yet a damage-limiting posture tends to contravene deterrence: one nation's measures to limit damage in nuclear war generate the perception by the opponent that such a nation would be more difficult to deter from initiating nuclear conflict. Similarly, extending doctrine to the use of nuclear weapons to deterring not only nuclear war but *war at all levels* projects the image that nuclear retaliation is threatened against even minor, nonnuclear incursions. Such a threat is hardly credible in the face of today's nuclear balance. As a matter of history, military intervention not using nuclear weapons either by nonnuclear nations or by the superpowers has not been deterred. The broadening of doctrine raises the perceived requirements for nuclear arms and thus contributes to the arms race.

On the Soviet side, the available literature is even less definite as to the priority given to deterrence in strategic doctrine; there are extensive discussions of extended fighting of nuclear war. Soviet leaders have made it clear that they are well acquainted with the potential horrors of nuclear war, and both Soviet and U.S. leaders continue to express great doubts about whether a limited nuclear war could ever remain limited. Yet the actual development and deployment of hardware on both sides can only have a "rational" basis if its use is anticipated in connection with various limited conflicts and if limited conflicts can remain confined.

Recently, in the United States, *limited* use of nuclear weapons has been emphasized more frequently in policy declarations. Part of the basis of this emphasis is "mirror imaging," citing selected Soviet sources on nuclear doctrine and Soviet weapons deployments. A "straw man" is presented: Under past doctrine, U.S. leaders would have no choice but to respond to the initiation of nuclear war of some kind either by not using nuclear weapons at all or by using them in a full-scale counterattack. This situation has never been real.

The so-called MAD (Mutual Assured Destruction) strategy has never existed as such. Rather, the option has always been available to launch parts of the force against a fraction of the listed targets, although the actual implementation of such "flexible" response can be—and is being—improved. Nevertheless, U.S. leaders, and particularly U.S. Secretary of Defense Harold Brown, continue to emphasize that they do not believe that a limited conflict will remain limited. At the same time, current doctrine is that the deterrent purpose of U.S. weapons is to convince an adversary that "first use" of nuclear weapons in any situation, for any purpose, limited or comprehensive, cannot lead to an outcome favorable or acceptable to the initiator. Rigidly interpreted, such a doctrine might imply that the United States should be prepared to fight to "victory" *any* nuclear war fought for any "limited" goal set by the opponent. More liberally interpreted, this doctrine deviates very little from earlier declarations on "flexible response."

Although currently stated U.S. doctrine implies, in fact, little significant change in numbers and kinds of weaponry, the frequent public reformulations of strategy give the false impression of truly significant changes in doctrine. This, in turn, gives the impression that consequently large changes in future military weapons may result from the changed doctrine, an impression that is largely false.

Once nuclear weapons are used at all in war, the ultimate impact is largely incalculable; therefore, the threat faced by mankind due to the advent and build-up of nuclear weapons is largely insensitive to announced military doctrine. This conclusion is based not only on the often-stated belief, which I share, that it is unlikely that "limited" war will stay limited but also on the actual physical and technical effects of so-called limited exchanges.

Most analyses of limited nuclear war tend to be grossly simplistic in largely ignoring the *collateral effects* of such exchanges. As an example, many studies on the use of nuclear weapons in theatre conflict do not include the effects of targeting errors or intelligence failures in identifying and locating targets. One must remember that, when a CEP (Circular Error Probable) is assigned to a weapons system, this means that one-half the rounds will impact outside this error circle, and some of those impact points just might be highly populated areas. Analyses also tend to ignore the fact that, in all past wars in Europe, a large fraction of the population—and, in particular, the urban population—did not remain in cities but traveled along roads to other areas. Yet calculations of refugee casualties are rarely included in analyses of theatre nuclear war. In addition, although some military doctrines discount the value of occupation of cities by troops in localized warfare, the taking of cities has been, in fact, a major political prize in past wars. Therefore, while the assumption is often made in analyses of limited warfare that enemy columns will avoid centers of population, this is likely to be false. On top of all this, civil-defense measures and evacuation, so much discussed publicly in the United States and implemented on a considerable scale in the Soviet Union, are likely to be highly ineffective, simplistic calculations to the contrary notwithstanding.

All these considerations indicate that the vast volume of discussions and literature on limited uses of nuclear weapons tends to disregard physical reality. National leaders should not assume that they can fine-tune the use of nuclear weapons in actual conflicts. This point can possibly best be illustrated by the so-called neutron bomb (or, more correctly, the enhanced-radiation warhead), which has played such a large *political* role in Europe. The neutron bomb is considered as a specialized device that minimizes blast while maximizing lethal effects from radiation. In this role, it is advertised to be a powerful weapon against the crews of enemy tanks, while minimizing collateral damage to friendly populations, structures, and economic assets of the country in which this device is used. Yet, under potential use of such a weapon, the actual physical differences between the neutron bomb and other nuclear weapons are minimal. Roughly speaking, the 1-kiloton version of the U.S. Lance Enhanced Radiation Warhead produces as much neutron radiation as would a "normal" 10-kiloton device. The blast effect is still equivalent to 1000 tons of TNT, much larger than any blockbuster of World War II. The lethal radius of the neutrons is significantly dependent on intervening structures, and there is a wide gap between a lethal radiation dose and an instantaneously disabling dose, the latter being about 20 times larger. All these physical factors, together with the problems cited above (that is, targeting errors, intelligence failures, and the presence of refugees in battle zones), indicate quite clearly that highly specialized use of enhanced-radiation warheads is not all that specialized. In other words, the enhanced radiation of the neutron bomb is unlikely to be exploitable.

The controversy over whether the neutron bomb will lower the nuclear threshold and thus make nuclear war more likely receives substance only if governments assume that the use of such a weapon implies consequences significantly different from the use of nuclear weapons of more usual design. In view of the previous discussion, such a conclusion is fallacious.

Defense Against Nuclear Weapons

The escalation in numbers of strategic nuclear weapons by the superpowers and the decrease in stability of the strategic deterrent due to the vulnerability of land-based missile silos to accurate enemy MIRVs have revived discussion of the need for widely or locally deployed antimissile defenses. The Soviet Union has carried out more extensive recent development programs than has the United States, and the Moscow ABM system is in the process of being modernized. While current activities are consonant with the provision of the SALT I Treaty, expanded deployments beyond these limits would invalidate the most significant arms-control agreement yet enacted by the Soviet Union and the United States.

One firm technical conclusion in the nuclear age is that the outcome of nuclear military conflict is apt to be "offense-dominated" rather than "defense-dominated." Since a single penetrating and exploding nuclear warhead does such an enormous amount of damage, the standards of interception required for defensive systems to be effective must be very high. Formal exchange calculations that compare the cost of defenses against nuclear weapons—be they air defenses or ballistic-missile defenses—with the incremental cost required for the offense to cancel the effect of the defense always show that defensive costs are higher if such calculations are carried out at relatively moderate levels of casualties. There could be specialized exceptions to this general conclusion. For example, defense of such hard-point targets as missile silos may, under certain circumstances, be militarily effective. However, the conclusion remains valid that, at least in reference to all currently designed systems, ABM is a poor investment for any large-scale strategic exchange, whether the targets are economic or military, and is an escalatory component of the arms race.

This negative *technical* conclusion on the value of antinuclear defenses is *politically* difficult to accept. As a matter of psychological and political necessity, it is unpalatable to leave the homeland undefended; yet the relative impotence of defense is a de facto truth in the nuclear age. Once there is clear recognition that nuclear strategic balance is offense-dominated, an escalating and self-defeating offense–defense competition can be avoided.

"Sufficiency"

All studies continue to show that, if a large fraction of the 50,000 warheads are now stockpiled worldwide—most are more powerful than the two that together killed 250,000 people in Japan—were used in war, several hundred million people would die and immense suffering would follow. Such studies make a variety of assumptions about targets, shelter, or effectiveness of evacuation. Yet they tend to consider only the "prompt" effects of nuclear weapons—blast and early radiation, combined with radioactive fallout. The casualties induced by delayed consequences—the effect of fires, food deficiency, absence or maldistribution of medical care, societal breakdown, epidemics—are not included; they are omitted as "too difficult to calculate." Some of the long-range effects—ecological imbalances, depletion of the ozone layer, synergistic effects, and genetic burdens—have been studied, but it is generally agreed that the unknown exceeds the known.

How, in the face of these horrendous facts, has a series of "rational" decisions led to the status quo? What reasoning has led the superpowers to conclude that their stockpiles of nuclear weapons and at times even their supplies of weapons-grade nuclear raw materials are insufficient? In the interest of the future of mankind, the kinds of practices and forces that have driven the armaments in the world to their current level must not continue.

Since most of the factors that drive the arms race are fundamentally not directly related to science and technology, remedies must be sought largely in the political arena. I would like to defend here the thesis that, if the technical and scientific nature of weaponry—and particularly of nuclear weaponry—were more clearly recognized and considered by the political decision-makers, then the largely nontechnical factors that now drive the arms race would be greatly diminished.

After the first nuclear weapon was detonated in 1945 at Alamogordo, many observers expressed the sentiment that this new weapon would make further wars impossible, and that further buildup of stockpiles of these weapons would not occur. These prophecies have proven false. *No strategic doctrine adopted officially by any power has provided a quantitative interpretation of the doctrine designated by Kissinger by the term "sufficiency." No government has answered the question, When is enough enough? applied to nuclear weapons.*

In principle, if there were an accepted answer to the quest for an interpretation of sufficiency, then limitation of nuclear arms could be achieved *both* through negotiated arms control *and* through unilateral action. Yet we have not seen a halt in the growth of nuclear stockpiles or in the number of deployed nuclear weapons. Fifty thousand nuclear weapons worldwide are still not "sufficient." I attribute this failure largely to a disregard by governmental leaders of generally-accepted technical facts about nuclear weapons.

Too often the radical difference between the nature of nuclear weapons and that of nonnuclear arms is ignored, or at least minimized by some. Conventional wisdom applied to nonnuclear weapons is that you require many more bullets than targets. For ordinary "bullets," this statement is self-evident, since a large fraction of available bullets would, in fact, not actually be fired in war; of those actually used, a large fraction would miss their intended targets. However, as applied to nuclear weapons, this notion should be far less persuasive. A nuclear weapon that misses its assigned target still does an enormous amount of damage. Since the principal aim of nuclear weapons, acknowledged by all parties, is to *deter* nuclear war, the importance of stockpiled weapons beyond those that might actually be used in a conflict is a matter of *political* leverage rather than *military* utility. *It is this political role of nuclear weapons, rather than their actual potential utility in warfare, that has prevented the nations of the world from defining nuclear "sufficiency."*

Conclusion

To summarize: *Once nuclear war is initiated by any power, under any doctrine, in any theatre, or for any strategic or tactical purpose, the outcome will result in truly massive casualties and devastation, leading to incalculable effects on the future of mankind. The predictions of science concerning the effects on large segments of the world's population of large-scale use of nuclear weapons are extremely uncertain.* If these facts were emphasized more often and more persuasively, then world leaders might begin to realize that such controversies as employment or nonemployment of such specialized weapons as the neutron bomb, or the exact numbers of weapons systems permitted under arms-control agreements, lack technical or military significance.

This discussion leads to the profoundly pessimistic conclusion that arms-control agreements have not been successful in limiting the destructive options of war—in particular, those involving nuclear weapons—since so many other forces and processes have, in fact, amplified the nuclear arms race. This trend must be reversed, or else the future will indeed be dim.

Many remedies for this condition have been discussed and will continue to be discussed. Leading the list of remedies must be a conscientious policy decision by the nuclear nations to govern their policies in regard to nuclear weapons by true technical and military considerations, and to minimize their justification as symbols of prestige and power.

More specific measures following from such a policy might be:

1. *Sufficiency.* Each nuclear nation should analyze its nuclear-weapons needs under a strategy solely dedicated to deter initiation of nuclear attacks by others under a variety of assumptions. Such deterrence should retain the option to counterattack a spectrum of military and economic targets with varying degrees of severity, but *not* to reply *in kind* to all possible modes of attack the initiator of nuclear war might choose. Sites that are of limited economic or military value, and those targets that cannot be attacked successfully, should be eliminated from the list of possible objectives of nuclear attack. This prescription should result in specifying the number of survivable nuclear weapons "sufficient" for national security, since, in the technical sense, the utility of nuclear weapons ceases to increase once certain numbers are reached.

2. *Arms Control.* Concurrent with such reexamination of policy and a drastically downward revision of the required number and kinds of weapons, arms-control negotiations must be given higher priority than they have been given in the past—or even the highest priority—in the discussion among nations. This implies that the linkage between arms-control issues and other topics of controversy between nations must be minimized. Those arms-control measures that improve the security of *both* sides should be emphasized.

3. *Unilateral Action.* Based on internal examination, each nuclear nation should consider unilateral steps in decreasing nuclear armament. Among the possibilities for such unilateral steps would be a cut-off of, or a reduction in, the production of weapons-grade, fissionable material, and the elimination of clearly redundant weapons and those weapons susceptible to first-strike attack (and therefore only useful for a first strike in themselves).

4. *The Technical Nature of Nuclear Weapons.* Since the current generation of decision-makers in most countries lacks personal acquaintance with nuclear weapons and their effects, a renewed effort should be made, within each country and internationally, to disseminate fully and publicly the technical facts about the nature of nuclear weapons and their destructive effects. Such studies should emphasize not only what is known but also what is unknown. International communication and exchanges among scientists, including those who have contacts with military activities, should be greatly amplified.

The hope that measures such as those listed here might lead to a halt, or even reversal, in the growing accumulation of nuclear arms is based on the conclusion that the nuclear arms race is driven by the *perception* of power rather than by physical realities. Unfortunately, perceptions tend to be *political* realities, and thus nuclear weapons become symbols of power and strength and bargaining chips in arms-control negotiations. Such a shift of the role of nuclear weapons—from weapons of mass destruction to political tools—has denied each nation a rationale for defining what nuclear weaponry is sufficient for its needs. Nuclear weapons cannot compensate for deficiencies in conventional military power. *Only a continuing and insistent reminder about the technical realities can form the basis for agreements and unilateral acts leading to lowered limits on nuclear weapons deemed sufficient for national and international security. The present course is set for disaster.*

The Mutual-Hostage Relationship between America and Russia

For nearly two decades, the strategic nuclear armaments of the Soviet Union and the United States have been great enough for each to hold the other's civilian population hostage against a devastating nuclear attack. Living with this situation has not been, and will not be, easy: it has become, quite simply, one of the major tensions of modern life. Yet the mutual-hostage relationship has been given credit, and probably justly so, for the prevention of massive world wars.

During the last few years, this relationship has been exposed to broader public scrutiny as a result of the SALT I negotiations and treaty, and a number of articles and statements have appeared criticizing U.S. Policy with regard to the situation.[1] One critic, Donald Brennan, coined the acronym MAD, for Mutual Assured Destruction, to indicate his view of the policy underlying SALT. While others have not employed terms quite as harsh, they still assert that the terms of the SALT I treaty prohibiting extensive antiballistic missile (ABM) deployments do in fact

[1] See, for instance: Michael May, *Orbis*, Summer 1970, pp. 271 ff., and Princeton Center of International Studies Research Monograph 37,1972; W. R. Van Cleave, *Freedom at Issue*, May–June 1973, No. 19; D. G. Brennan, "The Case for Missile Defense": *Foreign Affairs*, April 1969, and *Survival*, September–October 1972; Fred Charles Iklé, "Can Nuclear Deterrence Last Out the Century?" *Foreign Affairs*, January 1973 (a more complete edition of that article appeared as a report of the California Arms Control and Foreign Policy Seminar as of January 1973).

First published in *Foreign Affairs*, No. 54, 109–118 (1973). Reprinted by permission.

signify a morally repugnant policy of leaving "mass slaughter" as the only option in case deterrence has failed in some way.

The recently named head of the Arms Control and Disarmament Agency, Fred Charles Iklé, cites three "far-reaching dogmas" as implied by current U.S. policies:

> One: our nuclear forces must be designed almost exclusively for "retaliation" in response to a Soviet Nuclear attack—particularly an attempt to disarm us through a sudden strike.
>
> Two: our forces must be designed and operated in such a way that this retaliation can be swift, inflicted through a single, massive, and above all prompt strike. What would happen after this strike is of little concern for strategic planning.
>
> Three: the threatened "retaliation" must be the killing of a major fraction of the Soviet population; moreover, the same ability to kill our population must be guaranteed the Soviet government in order to eliminate its main incentive for increasing Soviet forces. Thus, deterrence is "stabilized" by keeping it mutual.[2]

The first of these "dogmas" conforms to the technical realities: in the post–SALT I era (and under conditions prevailing throughout the past decade!), our strategic forces must be designed primarily for retaliation in response to nuclear attack. However, I take strong exception to the second and third points, which claim that such a response, according to accepted doctrine underlying SALT, must be both rapid and of massive proportion.

Naturally, the present situation is far from ideal. We cannot be relieved of moral responsibility for having permitted a situation to develop in which large segments of the population of both West and East can, in fact, be sacrificed at the will of political leaders; neither is the situation free from acute danger in case of failure of mutual deterrence. Iklé aptly criticizes the mutual-hostage relationship that these policies imply by eloquently recalling that the threat of the killing of civilians has been condemned as immoral in the codes of both ancient and modern warfare, and also by pointing out the fragility of "stability through deterrence"—for example, in scenarios of accident and unauthorized nuclear attack.

---

[2] Iklé, *op. cit.*

Yet how can we do better? The critics seem to imply that the mutual-hostage relationship between the populations of the United States and the Soviet Union is a consequence of policy and would, therefore, be subject to change if such a policy were modified. Yet this relationship is a matter of physical fact and is thus grossly insensitive to any change in strategic policy. The reason is simple: the destructiveness of today's offensive arsenal of nuclear weapons is so overwhelming that deaths would number in the many millions, or even tens of millions, if only a fraction of the available weapons were delivered against the opponent's homeland.

In the face of this physical reality, much of the recent criticism has concentrated on extending a hope of "low-casualty" nuclear war by advocating a policy of strictly antimilitary attacks, or of "controlled" nuclear attacks against selected targets only, either military or civilian.

Neither of these scenarios, however, solves the basic problem of the mutual-hostage relationship. For no one can be sure whether an opponent will, in fact follow a similarly "restrained" policy; he may instead choose a full antipopulation response. Moreover, civilian casualties as a result of any massive antimilitary attacks would still be enormous. Finally, once the barriers against use of nuclear weapons are broken, escalation toward full-scale nuclear war is exceedingly difficult to prevent.

It is characteristic that none of the recent analyses meet these questions head-on. In essence, these papers start with the premise that "there must be a better way" than mutual deterrence, but the viability, let alone the advantage, of other specific policies is not demonstrated.

If the only defect of the criticism deploring the state of mutual deterrence now extant between the United States and the Soviet Union were a failure to provide concrete alternative prescriptions that would be more likely to prevent nuclear war, this would be a matter of little concern. However, the problem is deeper. Any successful attempt to project an image—however ill-founded—of a "clean" nuclear war generating minimum civilian casualties could make the use of nuclear weapons in limited conflicts more acceptable. The fact remains—irrespective of the extent to which the strategies of either country include plans for deliberate retaliation against the opponent's population—that the peoples of both countries are in jeopardy in any kind of nuclear conflict.

This is not the first time these question have been raised—far from it. After former Secretary of Defense McNamara took office, he proclaimed, in 1962, a "city-avoiding" strategy designed to minimize civilian casualties in a nuclear war. But, during his tenure in office, he gradually became convinced that such a policy was unworkable, both on physical and on military grounds: civilian casualties in connection with a purely antimilitary attack were still likely to number in the many millions,[3] and one could not be sure the opponent would also follow a city-avoiding strategy; instead, he might choose to target centers of population! Accordingly, McNamara, in his later years, completely changed his position, moving toward a policy of deterrence and, more specifically, "assured destruction." Without going into the merit of McNamara's conversion, this history illustrates that, throughout a period of major change in strategic policy, the mutual-hostage relationship between the U.S. and Soviet populations remained a physical reality of central importance.

In the face of the unavoidable fact that the populations of both countries are exposed to overwhelming danger in case of nuclear war, emphasis has been placed on stabilizing the political, economic, and military relationships between the two countries. In the purely military-strategic sense, this search for stability has taken the form of an effort to remove any *rational* incentive for the initiation of nuclear war. In principle, such an incentive could be removed if technology permitted the evolution of active and passive defense measures that would either almost totally prevent the arrival of enemy nuclear weapons or eliminate their devastating consequences. In view of the enormous destructiveness of each penetrating warhead and the low performance and high cost of feasible defensive measures, a state of defense-dominated stability appears unattainable. (Brennan remains the principal dissenter from this conclusion.) A defense deployed against a massive nuclear attack, using any technology now known or surmised, would be enormously expensive if it were designed to hold casualties to a small percentage of the population; moreover, a relatively less expensive increase in the opponents' offensive forces would cancel the protection provided by such a defense. These conclusions are valid even if highly optimistic assumptions are made regarding the performance of defense measures—which can never realistically be tested.

Once a tight defensive umbrella is ruled out, stability rests on deterrence achieved by protection of the strategic offensive forces against a totally disabling pre-emptive first strike. This, in turn, implies that the stra-

---

[3] Calculations indicate that an attack against all of the U.S. Minuteman silos would result in casualties (U.S. or Canadian) in the multimillion range from fallout alone; the actual numbers from all causes would be still larger.

tegic nuclear weapons on either side must be protected against initial attack through hardening, mobility, or secrecy of locations, and that any moves *on either side* that would impair these values would be considered destabilizing. Thus, in this situation, the deployment of weapons suitable for an effective attack against the strategic retaliatory forces of the opponent (often termed a "counterforce attack") detract from the margin of stability. However, stability, as it has now been achieved, does not imply that there is only one preordained option with which the strategic forces would retaliate if they were subject to attack. It is this latter point that apparently is frequently misunderstood by critics of the present situation.

What really are the choices that our strategic systems permit in the event the country is attacked? Once the technical nature of the forces is restricted, making them unsuitable for an effective first strike against the other side's strategic weapons, then certain types of counterforce responses to an enemy attack should no longer be considered real options. Thus, any response that could be interpreted as part of a first-strike posture is ruled out. Specifically, a counterforce attack—which, by virtue of its explosive power, accuracy, and number of warheads might endanger the land-based, hardened silos of the other side—would have to be explicitly ruled out because that same configuration would also be an essential component of a first-strike attack. However, this is the *only* restriction on the procedure that is not imposed by deliberate policy or by technical conditions subject to modification. This means that, while I agree with Iklé's first point, "that our nuclear forces must be designed almost exclusively for retaliation," as correctly representing the cornerstone of stability in this epoch, the nature of such retaliation in given wide latitude; the other two points made by Iklé in his article seriously misrepresent the current situation by claiming that response is restricted to instant and massive retaliation.

In fact, there is no basic *technical* reason why any retaliation would have to be swift; a great deal of technical, political, and diplomatic effort during the last two decades has gone into measures to prevent just that compulsion. The "hot line" that was first established by the Memorandum of Understanding of June 20, 1963, and whose character was upgraded during the SALT talks, is a case in point. The very purpose of that hot line is to permit an exchange of information among the parties in case a nuclear explosion has taken place on the territory of one of them, and this communication does not preclude subsequent retaliation. Similarly, efforts have been made to harden and diversify command-and-control systems so that no instantaneous "go" orders must be sent out on first verification of nuclear attack, let alone on warning. Whether these measures on either side of the ocean are fully adequate technically is, of course, a matter about which residual doubt will always remain. Increased awareness of this problem and measures to improve the situation technically are certainly needed.

Those measures (such as improving the accuracy of strategic warheads, adding multiple warheads to intercontinental missiles, and so on) that appear to threaten land-based missile silos are the main causes of arguments (albeit highly unpersuasive) supporting the need for rapid response, or even advocating the tactic of launching missiles on receipt of warning but before actual impact of energy weapons (designated "launch-on-warning"). There has been a flood of calculations regarding the alleged vulnerability of the land-based Minuteman forces. Such projections use a range of numbers of enemy missiles that assume multiple Soviet warheads yet to be developed; these are to impact with assumed explosive powers and at various distances from the silos. Calculations on the survival of strategic aircraft under attack and estimates of their ability to deliver nuclear weapons are less numerous and more difficult; there are no meaningful calculations at all on the vulnerability of our strategic submarines.

Apparently, the reason for this disparity of calculational effort is that computations of Minuteman vulnerability can easily be made with a wide variety of models, even if the assumptions are difficult to justify; there is no specifically known physical vulnerability for nuclear missile submarines. However, calculations even for Minuteman are dubious, since Minuteman vulnerability is very steeply dependent on the accuracy of the attacking missiles and the performance of such missiles and hardened silos under actual combat conditions is uncertain. Moreover, it is very difficult to predict precisely how, in a heavy attack, one missile will affect another: dust or debris produced by one missile impact may destroy another incoming warhead, and the radiation from one nuclear explosion can disable a companion missile. For these and other reasons, an attacker could have little confidence in his calculated ability to reduce the number of Minuteman survivors to the very small number "negligible" as a retali-

atory threat.[4] Thus, even though the more extreme projections of the Soviet threat beyond this decade (unless limited by future SALT agreements) indicate very few Minuteman survivors from a Soviet attack, Minuteman remains a substantial contributor to deterrence.

Whatever the vulnerability may be of each member of the "triad" of
strategic retaliatory forces (submarines, land-based missiles, and bombers), there is no technical method in view by which either side could
mount a fully disabling and synchronized attack against the *combination*
of nuclear strategic forces of the other. Thus, neither the present nor the
foreseeable technical situation creates a need for a rapid, and possibly ill-
considered, response to attack.

It is equally incorrect to state that such a counterattack must be a single
massive strike. With the exception of being denied a counterforce strike
against the other side's hardened silos, the choice in numbers and kinds of
targets—be they military or civilian—is governed only by the technical
features of the command-and-control system and the doctrine that governs its application. Therefore, the answer to President Nixon's oft-quoted
question: "Should a president, in the event of a nuclear attack, be left with
the single option of ordering the mass destruction of enemy civilians, in
the face of the certainty that it would be followed by the mass slaughter
of Americans?" (delivered as part of the State of the World message in
1970) is, "No, he should not." And he does indeed have many other
choices. Thus, Iklé's third point—"The threatened 'retaliation' must be
the killing of a major fraction of the Soviet population"—also does not
correctly describe the current situation, either before SALT or after.

President Nixon himself has now said, in his fourth foreign policy
message of May 1973: "An aggressor, in the unlikely event of nuclear
war, might choose to employ nuclear weapons selectively and in limited
numbers for limited objectives," and, "If the United States has the ability

---

[4] A retaliatory attack would have to be very small indeed to be "negligible." Indeed, neither
leaders nor serious observers in either country should pay much attention to the spuriously
precise analyses cranked out by military computers to "determine" levels of damage from
nuclear attack. Such calculations usually take into account only "prompt" casualties—that is,
those resulting from blast or prompt radiation. Few analyses consider fallout, and none of those
generally used take into account such postattack effects as fire, damage to food supplies,
medical care and productivity, or epidemics. As Iklé notes, the omission of such after-attack
effects leads to substantial underestimates. It is another instance of the way in which, in his
words: "The jargon of American strategic analysis works like a narcotic. It dulls our sense of
moral outrage about the tragic confrontation of nuclear arsenals, primed and constantly
perfected to unleash widespread genocide."

to use its forces in a controlled way, the likelihood of nuclear response would be more credible, thereby making deterrence more effective and the initial use of nuclear weapons by an opponent less likely." These statements justify more convincingly the need for a large variety of nuclear options—not as a means to abolish the mutual-hostage relationship between U.S. and Soviet citizens but to strengthen deterrence against first use of nuclear weapons of *all* kinds. SALT has not impaired these more limited responses; on the contrary, the severe ABM restraints of the SALT treaty have assured penetration of even small missile attacks and therefore have broadened the range of possible retaliation.

Mankind has indeed succeeded in creating a situation in which the vast stockpiles of nuclear weapons in the world can no longer be "rationally" used. But is that enough? Although the preceeding discussion clearly refutes the claim of Iklé and others that the present strategic doctrine requires a rapid and massive retaliatory response, the critics have performed a valuable service by shaking confidence in the long-range "stability" that the present arrangements imply. Whether or not credit has been given correctly to the role of nuclear weapons in having prevented large-scale war after World War II, it is true that this record may be broken at any time by a nuclear accident, by escalation of a war initiated by third powers, or by unauthorized attacks. There is no meaningful way to predict whether these "irrational" nuclear catastrophes can be avoided throughout this century and beyond as long as the enormous nuclear stockpiles grow, or even remain.

On the positive side, there is increasing pressure for more layers of safety devices, better communications, and so on. Moreover, there may also be hope that Permissive Actions Links (PALS)—devices that, by mechanical means, prevent one military echelon from executing a strike without permission from a higher level—may be used for strategic as well as tactical nuclear-weapons systems. On the negative side, we have the ever-increasing complexity of nuclear delivery systems and the increasing destructive power at the command of a single submarine commander. Finally, there is the problem of maintaining high standards of diligence and responsibility on a routine basis for a large number of years.

A possible constructive step in arms-control negotiations would be an agreement on progressively tightening the political and technical command-and-control provisions over the strategic systems of the nuclear powers. This is clearly a move not subject to verification, but the incentive to violate such an accord appears small enough that such a provision might be negotiable.

In the last analysis, however, the risk of accidental war cannot be eliminated. Our hope for avoiding a nuclear catastrophe over the long range rests on continually reducing the product of the two variables that define the risk—the number of nuclear weapons in strategic stockpiles, and the chance of any one of them being delivered through accidental launch or unauthorized use. Without a steady decrease in this index, the future is indeed dim.

In short, even though the present degree of stability is greater than the critics suggest, there can be no assurance that it will, in fact, prevent the outbreak of nuclear war either by accident or through conflict introduced by third countries. The critics of the present doctrine have done substantial harm by their unsubstantiated claim that some strategic policy—not accompanied by a dramatic reversal in the growth of nuclear armaments—can relieve the inhumanity of the present situation, even perhaps the risk of accidental war.

In another respect, the emphasis of the SALT critics on the use of nuclear weapons against military targets has given new incentives and justification for the procurement of counterforce weapons, such as highly accurate nuclear warheads. Such developments would be destabilizing by being physically indistinguishable from weapons designed for a pre-emptive attack against the opponent's retaliatory forces. In addition, there is the revival of the word "controlled." This refers to the military use of strategic nuclear weapons in the fighting of actual wars, while presumably minimizing the risk of escalation to a full-scale nuclear conflict. Yet, if such a risk could really be minimized—a highly dubious assumption—then such a development would, in fact, remove a factor that now deters the outbreak of large-scale war.

I do not know or foresee a solution to the problem Iklé states: "By taking advantage of modern technology, we should be able to escape the evil dilemma that the strategic forces on both sides must either be designed to kill people or else jeopardize the opponent's confidence in his deterrent." In the absence of any specifically proposed (let alone established) resolution of this problem, statements such as these tend to mislead civilian policy makers and extend false hopes that technology will lead us out of the nuclear dilemma. Ill-founded attempts to "sanitize" nuclear war are a disservice to the maintenance of stability, as well as to efforts to reduce areas of risk.

In essence, the critics of a primarily deterrent posture and the advocates of actually fighting nuclear wars assume that scientific progress will somehow alter the existing realities. I can see no technological basis for this assumption. Specifically:

- No technological distinction exists or can be created between those nuclear weapons endangering the deterrent forces of the opponent in a first or pre-emptive strike (and thus decreasing stability) and weapons designed to attack the same forces by retaliation.
- There is no demonstrable break between nuclear weapons designed for limited attacks and those designed for "strategic" retaliation.
- Antimilitary nuclear attacks of substantial size will almost certainly generate enormous civilian casualties.
- Whatever plans or technological preparations the United States may make to fight a "controlled" nuclear conflict, there can be no certain method to protect the U.S. population in case the opponent decides to respond with an antipopulation attack.
- Available casualty estimates understate the effects of large-scale nuclear war; such consequences as epidemics aggravated by maldistribution of medical care, fire, starvation, ecological damage, and societal breakdown are well-nigh incalculable.

From these inescapable conditions it follows, in my judgment, that the only clear demarcation line giving a "fire-break" in the use of weapons in war will continue to be the boundary between nonnuclear and nuclear devices. Mere shifts in policy and strategic doctrine will neither eliminate the hostage role of the populations of the United States and the Soviet Union nor decrease the danger of nuclear catastrophe through accident or through unauthorized attack. Nor will they, in Churchill's words, "cover the case of lunatics or dictators in the mood of Hitler when he found himself in his final dugout." Only the relaxation of political tensions, coupled with bold steps limiting and reducing the quality and quantity of arms, and with ever-increasing vigilance over the control, safety, and nonproliferation of nuclear weapons, can offer hope that nuclear disaster can be avoided.

MAD versus NUTS

WITH SPURGEON M. KEENY, JR.

S ince World War II, there has been a continuing debate on military doctrine concerning the actual utility of nuclear weapons in war. This debate, irrespective of the merits of the divergent points of view, tends to create the perception that the outcome and scale of a nuclear conflict could be controlled by the doctrine or the types of nuclear weapons employed. Is this the case?

We believe not. In reality, the unprecedented risks of nuclear conflict are largely independent of doctrine or its application. The principal danger of doctrines that are directed at limiting nuclear conflicts is that they might be believed and form the basis for action without appreciation of the physical facts and uncertainties of nuclear conflict. The failure of policy-makers to understand the truly revolutionary nature of nuclear weapons as instruments of war and the staggering size of the nuclear stockpiles of the United States and the Soviet Union could have catastrophic consequences for the entire world.

Since the end of World War II, military planners and strategic thinkers have sought ways to apply the tremendous power of nuclear weapons against target systems that might contribute to the winning of a future war. In fact, as long as the United States held a virtual nuclear monopoly, the targeting of atomic weapons was looked upon essentially as a more effective extension of the strategic bombing concepts of World War II. With the advent in the mid-1950s of a substantial Soviet nuclear capabil-

First published in *Foreign Affairs*, No. 60, 287–304 (1981). Reprinted by permission.

ity, including multimegaton thermonuclear weapons, it was soon apparent that the populations and societies of both the United States and the Soviet Union were mutual hostages. A portion of the nuclear stockpile of either side could inflict on the other as many as 100 million fatalities and destroy it as a functioning society. Thus, although the rhetoric of declaratory strategic doctrine has changed over the years, mutual deterrence has in fact remained the central fact of the strategic relationship of the two superpowers and of the NATO and Warsaw Pact alliances.

Most observers would agree that a major conflict between the two hostile blocs on a worldwide scale during this period may well have been prevented by the specter of catastrophic nuclear war. At the same time, few would argue that this state of mutual deterrence is a very reassuring foundation on which to build world peace. In the 1960s, the perception of the basic strategic relationship of mutual deterrence came to be characterized as "Mutual Assured Destruction," which critics were quick to note could be reduced to the acronym MAD. The notion of MAD has been frequently attacked not only as militarily unacceptable but also as immoral, since it holds the entire civilian populations of both countries as hostages.[1]

As an alternative to MAD, critics and strategic innovators have, over the years, sought to develop various doctrines that would somehow retain the use of nuclear weapons on the battlefield or even in controlled strategic-war scenarios, while sparing the general civilian population from the devastating consequences of nuclear war. Other critics have found an alternative in a defense-oriented military posture designed to defend the civilian population against the consequences of nuclear war.

These concepts are clearly interrelated, since such a defense-oriented strategy would also make a doctrine including the actual fighting of nuclear wars more credible. But both alternatives depend on the solution of staggering technical problems. A defense-oriented military posture requires a nearly impenetrable air and missile defense over a large portion of the population. And any attempt to have a controlled war-fighting capability during a nuclear exchange places tremendous requirements not only on decisions made under incredible pressure by men in senior positions of responsibility but on the technical performance of command, control, communications, and intelligence functions—called in profes-

---

[1] See, for example, Fred Charles Iklé, "Can Nuclear Deterrence Last out the Century?", *Foreign Affairs*, January 1973, pp. 267–85.

sional circles "C³I" and which, for the sake of simplicity, we shall hereafter describe as "control mechanisms." It is not sufficient as the basis for defense policy to assert that science will "somehow" find solutions to critical technical problems on which the policy is dependent, when technical solutions are nowhere in sight.

In considering these doctrinal issues, it should be recognized that there tends to be a very major gap between declaratory policy and actual implementation expressed as targeting doctrine. Whatever the declaratory policy might be, those responsible for the strategic forces must generate real target lists and develop procedures under which various combinations of targets could be attacked. In consequence, the perceived need to attack every listed target, even after absorbing the worst imaginable first strike from the adversary, creates procurement "requirements," even though the military or economic importance of many of the targets is small.

In fact, it is not at all clear in the real world of war planning whether declaratory doctrine has generated requirements or whether the availability of weapons for targeting has created doctrine. With an estimated 30,000 warheads at the disposal of the United States, including more than 10,000 avowed to be strategic in character, it is necessary to target redundantly all urban areas and economic targets and to cover a wide range of military targets in order to frame uses for the stockpile. And, once one tries to deal with elusive mobile and secondary military targets, one can always make a case for requirements for more weapons and for more specialized weapon designs.

These doctrinal considerations, combined with the superabundance of nuclear weapons, have led to a conceptual approach to nuclear war that can be described as Nuclear Utilization Target Selection. For convenience, and not in any spirit of trading epithets, we have chosen the acronym NUTS to characterize the various doctrines that seek to utilize nuclear weapons against specific targets in a complex of nuclear-war situations intended to be limited, as well as the management over an extended period of a general nuclear war between the superpowers.[2]

While some elements of NUTS may be involved in extending the

---

[2] The acronym NUT for Nuclear Utilization Theory was used by Howard Margolis and Jack Ruina, "SALT II: Notes on Shadow and Substance," *Technology Review*, October 1979, pp. 31–41. We prefer Nuclear Utilization Target Selection, which relates the line of thinking more closely to the operational problem of target selection. Readers not familiar with colloquial American usage may need to be told that "nuts" is an adjective meaning "crazy or demented." For everyday purposes, it is a synonym for "mad."

credibility for our nuclear deterrent, this consideration in no way changes the fact that mutual assured destruction, or MAD, is inherent in the existence of large numbers of nuclear weapons in the real world. In promulgating the doctrine of "countervailing strategy" in the summer of 1980, President Carter's Secretary of Defense Harold Brown called for a buildup of nuclear war-fighting capability in order to provide greater deterrence by demonstrating the ability of the United States to respond in a credible fashion without having to escalate immediately to all-out nuclear war. He was very careful, however, to note that he thought that it was "very likely" that the use of nuclear weapons by the superpowers at any level would escalate into general nuclear war.[3] This situation is not peculiar to present force structures or technologies; and, regardless of future technical developments, it will persist as long as substantial nuclear-weapon stockpiles remain.

Despite its possible contribution to the deterrence of nuclear war, the NUTS approach to military doctrine and planning can very easily become a serious danger in itself. The availability of increasing numbers of nuclear weapons in a variety of designs and delivery packages at all levels of the military establishment inevitably encourages the illusion that, somehow, nuclear weapons can be applied in selected circumstances without unleashing a catastrophic series of consequences. As we shall see in more detail below, the recent, uninformed debate on the virtue of the so-called neutron bomb as a selective device to deal with tank attacks is a depressing case in point. NUTS creates its own endless pressure for expanded nuclear stockpiles with increasing danger of accidents, accidental use, diversions to terrorists, and so on. But, more fundamentally, it tends to obscure the fact that the nuclear world is, in fact, MAD.

The NUTS approach to nuclear war-fighting will not eliminate the essential MAD character of nuclear war for two basic reasons that are rooted in the nature of nuclear weapons and the practical limits of technology. First, the destructive power of nuclear weapons, individually and most certainly in the large numbers discussed for even specialized application, is so great that the collateral effects on persons and property would be enormous and, in scenarios that are seriously discussed, would be hard to distinguish from the onset of general nuclear war. But, more

---

[3] See Harold Brown, Speech at the Naval War College, August 20, 1980, the most authoritative public statement on the significance of Presidential Directive 59, which had been approved by President Carter shortly before.

fundamentally, it does not seem possible, even in the most specialized
utilization of nuclear weapons, to envisage any situation in which escalation to general nuclear war would probably not occur, given the dynamics
of the situation and the limits of the control mechanisms that could be
made available to manage a limited nuclear war. In the case of a pro-
tracted, general nuclear war, the control problem becomes completely un-
manageable. Finally, there does not appear to be any prospect, for the
foreseeable future, that technology will provide a secure shield behind
which the citizens of the two superpowers can safely observe the course
of a limited nuclear war on other people's territory.

So much has been said and written about the terrible consequences of
nuclear war that any brief characterization of the problem seems
strangely banal. Yet it is not clear how deeply the horror of such an event
has penetrated the public consciousness or even the thinking of knowledgeable policymakers who, in theory, have access to the relevant information. The lack of public response to authoritative estimates that general
nuclear war could result in 100 million fatalities in the United States suggests a general denial psychosis when the public is confronted with the
prospect of such an unimaginable catastrophe. It is interesting, however,
that there has been a considerable reaction to the campaign by medical
doctors in several countries (including the United States and the Soviet
Union) that calls attention to the hopeless plight of the tens of millions of
casualties who would die over an extended period due to the total inability of surviving medical personnel and facilities to cope with the situation. One can stoically ignore the inevitability of death, but the haunting
image of being among the injured survivors who would eventually die
unattended is a prospect that few can easily accept fatalistically.

It is worth repeating the oft stated, but little comprehended, fact that a
single modern strategic nuclear weapon could have a million times the
yield of the high-explosive strategic bombs of World War II, or one hundred to a thousand times the yield of the atomic bombs that destroyed Hiroshima and Nagasaki, killing 250,000 people. The blast from a single 1-
megaton weapon detonated over the White House in Washington, D.C.,
would destroy multistory concrete buildings out to a distance of about 3
miles (10 pounds per square inch overpressure, with winds of 300 miles
per hour)—a circle of almost complete destruction reaching the National
Cathedral to the northwest, the Kennedy Stadium to the east, and across
the National Airport to the south. Most people in this area would be killed

immediately. The thermal radiation from the same weapon would cause spontaneous ignition of clothing and household combustibles to a distance of about 5 miles (25 calories per square centimeter)—a circle of raging fires reaching out to the District line. Out to a distance of almost 9 miles, there would be severe damage to ordinary frame buildings and second-degree burns to exposed individuals. Beyond these immediate effects, the innumerable separate fires that had been ignited would either merge into an outward-moving conflagration or, more likely, create a giant fire storm of the type Hamburg and Tokyo experienced on a much smaller scale in World War II. While the inrushing winds would tend to limit the spread of the fire storm, the area with 5 to 6 miles of the explosion would be totally burned out, killing most of the people who might have escaped initial injury in shelters.

The point has been forcefully made recently by members of the medical community that the vast numbers of injured who escape death at the margin of this holocaust could expect little medical help. But, beyond this, if the fireball of the explosion touched ground, the resulting radioactive debris would produce fallout with lethal effects far beyond the site of the explosion. Assuming the prevailing westerly wind conditions, a typical fallout pattern would indicate that there would be levels of fallout greater than 1000 rems (450 rems produces 50 percent fatalities) over an area of some 500 square miles, and more than 100 rems (the level above which there will be significant health effects) over some 4000 square miles, reaching all the way to the Atlantic Ocean. In the case of a single explosion, the impact of the fallout would be secondary to the immediate weapons effects, but, when there are many explosions, the fallout becomes a major component of the threat, since the fallout effects from each weapon are additive, and the overlapping fallout patterns would soon cover large portions of the country with lethal levels of radiation.

Such levels of human and physical destruction are difficult for anyone, layman or specialist, to comprehend, even for a single city, but, when extended to an attack on an entire country, they become a dehumanized maze of statistics. Comparison with past natural disasters is of little value. Such events as dam breaks and earthquakes result in an island of destruction surrounded by sources of help and reconstruction. Nuclear war with many weapons fired would deny the possibility of relief by others.

When General David Jones, Chairman of the Joint Chiefs of Staff, was asked at a hearing of the Senate Foreign Relations Committee on November 3, 1981, what would be the consequences in the northern hemisphere of an all-out nuclear exchange, he had the following stark response:

> We have examined that over many, many years. There are many assumptions that you have as to where the weapons are targeted. Clearly, the casualties in the northern hemisphere could be, under the worst conditions, into the hundreds of millions of fatalities. It is not to the extent that there would be no life in the northern hemisphere, but if all weapons were targeted in such a way as to give maximum damage to urban and industrial areas, you are talking about the greatest catastrophe in history by many orders of magnitude.

A devastating attack on the urban societies of the United States and the Soviet Union would, in fact, require only a very small fraction of the more than 50,000 nuclear weapons currently in the arsenals of the two superpowers. The United States is commonly credited with having some 30,000 nuclear warheads, of which well over 10,000 are carried by strategic systems capable of hitting the Soviet Union. It is estimated that the Soviet Union will soon have some 10,000 warheads in its strategic forces capable of hitting the United States. An exchange of a few thousand of these weapons could kill most of the urban population and destroy most of the industry of both sides.

But such figures are, in themselves, misleading, because they are already high on a curve of diminishing returns, and much smaller attacks could have very severe consequences. A *single* Poseidon submarine captain could fire some 160 independently targetable nuclear warheads (each with a yield several times larger than those of the weapons that destroyed Hiroshima and Nagasaki) against as many Soviet cities. If optimally targeted against the Soviet population, this alone could inflict some 30 million fatalities. One clear fact of the present strategic relationship is that the urban societies of both the United States and the Soviet Union are completely vulnerable to even a small fraction of the other side's accumulated stockpile of nuclear weapons.

The theme that nuclear weapons can be successfully employed in warfighting roles, somehow shielded from the MAD world, appears to be recurring with increasing frequency and seriousness.[4] Support for Nuclear

---

[4] For a particularly clear statement of this view, see Colin S. Gray and Keith Payne, "Victory Is Possible," *Foreign Policy*, Summer 1980, pp. 14–27. For opposing arguments, see Michael E. Howard, "On Fighting a Nuclear War," *International Security*, Spring 1981, pp. 3–17, and a further exchange between Messrs. Gray and Howard in *International Security*, Summer 1981, pp. 185–87.

Utilization Target Selection—NUTS—comes from diverse sources: those who believe that nuclear weapons should be used selectively in anticipated hostilities; those who believe that such capabilities deter a wider range of aggressive Soviet acts; those who assert that we must duplicate an alleged Soviet interest in war-fighting; and those who are simply trying to carry out their military responsibilities in a more "rational" or cost-effective manner. The net effect of this increasing, publicized interest in NUTS is to obscure the almost inevitable link between any use of nuclear weapons and the grim "mutual hostage" realities of the MAD world. The two forces generating this link are the collateral damage associated with the use of nuclear weapons against selected targets and the pressures for escalation of the level of nuclear force, once it is used in conflict. Collateral effects and pressures for escalation are themselves closely linked.

To appreciate the significance of the collateral effects of nuclear weapons and the pressure for escalation, one must look at actual war-fighting scenarios that have been seriously proposed. The two scenarios that are most often considered are Soviet attempts to carry out a disarming, or partially disarming, attack against U.S. strategic forces in order to force the surrender of the United States without war, and the selective use of nuclear weapons by the United States in Western Europe to prevent the collapse of NATO forces in the face of an overwhelming Soviet conventional attack. One can expect to hear more about the selective use of nuclear weapons by the United States in the Middle East in the face of an overwhelming Soviet conventional attack on that area.

The much discussed "window of vulnerability" is based on the fear that Soviets might launch a "surgical" attack against vulnerable Minuteman ICBM silos—the land-based component of the U.S. strategic triad—to partially disarm the U.S. retaliatory forces, confident that the United States would not retaliate. The scenario then calls for the United States to capitulate to Soviet-dictated peace terms.

Simple arithmetic based on intelligence assessments of the accuracy and yields of the warheads on Soviet missiles and the estimated hardness of Minuteman silos does indeed show that a Soviet attack leaving only a relatively small number of surviving Minuteman ICBMs is mathematically possible in the near future. There is much valid controversy about whether such an attack is, in fact, operationally feasible with the confidence that a rational decision-maker would require. But what is significant here is the question of whether the vulnerability of Minuteman, real or perceived, could, in fact, be exploited by the Soviets without risking general nuclear war. Would a U.S. president react any differently in response to an attack against the Minuteman force than to an attack of comparable weight against other targets?

Despite the relatively isolated location of the Minuteman ICBM fields, there would be tremendous collateral damage from such an attack, which, under the mathematical scenario, would involve at least 2000 weapons with megaton yields. It has been estimated, by the Congressional Office of Technology Assessment, that such an attack would result in from 2 million to 20 million American fatalities, primarily from fallout, since at least half the weapons would probably be ground-burst to maximize the effect of the attack on the silos. The range of estimated fatalities reflects the inherent uncertainties in fallout calculations due to different assumptions on such factors as meteorological conditions, weapon yield and design, height of burst, and amount of protection available and used. Estimates of fatalities below 8 million require quite optimistic assumptions.

It seems incredible that any Soviet leader would count on any U.S. president suing for peace in circumstances in which some 10 million American citizens were doomed to a slow and cruel death but the United States still retained 75 percent of the strategic forces and its entire economic base. Instead, Soviet leadership would perceive a president confronted with an incoming missile attack of at least 2000 warheads and possibly many more to follow in minutes. They would anticipate such a president weighing options of retaliating on warning with his vulnerable land-based forces or riding out the attack and then retaliating at a level and manner of his own choosing, with substantial surviving air and sea-based strategic nuclear forces.

It is hard to imagine that this scenario would give the Soviets much confidence in their ability to control escalation of the conflict, to avoid vast damage to their homeland, and to assure the survival of their leadership. If the Soviets did not choose to attack the U.S. command, control, communications, and intelligence (C^3I) capabilities, the United States would clearly be in a position to retaliate massively or to launch a more selective initial response. If vulnerable control assets were concurrently attacked, selective responses might be jeopardized, but the possibility of an automatic massive response would be increased, since the nature of the attack would be unclear. But, even if these control assets were initially untouched, the Soviets could not be so overly confident of their own control mechanisms, or so overly impressed with those of the United

States, as to imagine that either system could long control such massive levels of violence, with increasing collateral damage, without the situation very rapidly degenerating into general nuclear war.

The question of nuclear war-fighting in Europe has a long and esoteric history. Tactical nuclear weapons have been considered an additional deterrent to a massive Soviet conventional attack by threatening escalation to general nuclear war employing strategic forces—the so-called coupling effect. At the same time, tactical nuclear forces have been looked on as a necessary counterbalance to superior Soviet conventional forces in a war limited to Europe. To this end, the United States is said to have some 6000 to 7000 tactical nuclear weapons in Europe.[5] The existence of this stockpile has been public knowledge so long that it is largely taken for granted, and the power of the weapons, which range in yield from around 1 kiloton to around 1 megaton, is not appreciated. It is interesting to note that, in Europe, we have one nuclear weapon (with an average yield probably comparable to the weapon that destroyed Hiroshima) for every 50 American soldiers stationed there, including support troops. Tactical nuclear weapons, are of course, no longer a U.S. monopoly. The Soviets are building up comparable forces, and they have had, for some time, long-range theater nuclear missiles, earlier the SS-4 and SS-5, and now the SS-20, for which the United States does not have a strict counterpart. In this regard, it must be remembered that it is always feasible for the United States or the Soviet Union to employ some of their long-range strategic missiles against targets in Europe.

There is now a great debate, particularly in Europe, about the proposed deployment on European soil of U.S.-controlled long-range Pershing II and ground-based cruise missiles capable of reaching the territory of the Soviet Union, in response to the growing deployment of Soviet SS-20 mobile medium-range ballistic missiles. This discussion tends to consider the SS-20s and the proposed new forces as a separate issue from the short- and medium-range nuclear weapons already deployed in Europe. There is indeed a technical difference: the proposed Pershing II missile is of sufficient range to reach Soviet territory in only a few minutes, and the SS-20 is a much more accurate and flexible weapon system than earlier Soviet intermediate-range nuclear systems. Yet, the overriding issue,

---

[5] For a discussion of the usefulness of theater nuclear forces in NATO as of that date, see Alain C. Enthoven, "U.S. Forces in Europe: How Many? Doing What?", *Foreign Affairs*, April 1975, pp. 523–31.

which tends to be submerged in the current debate, is the fact that any use of nuclear weapons in theater warfare in Europe would almost certainly lead to massive civilian casualties, even in the unlikely event that the conflict did not escalate to involve the homelands of the two superpowers.

Calculations of collateral casualties accompanying nuclear warfare in Europe tend to be simplistic in the extreme. First, the likely proximity to the combat zone of highly populated areas must be taken into account. One simply cannot assume that invading enemy columns will position themselves so that they offer the most favorable isolated target to nuclear attack. Populated areas could not remain isolated from the battle. Cities would need be defended or they would become safe stepping-stones for the enemy's advance. In either case, it is difficult to imagine cities and populated areas remaining sanctuaries in the midst of a tactical nuclear war raging around them. Then one must remember that, during past wars in Europe, as much as one-half of the population was on the road in the form of masses of refugees. Above all, in the confusion of battle, there is no control system that could assure that weapons would not inadvertently strike populated areas. Beyond immediate effects, nuclear fallout would not recognize restrictions based on population density.

The common feature of the preceding examples is that specialized use of nuclear weapons will, as a practical matter, be difficult to distinguish from unselective use in the chaos of tactical warfare. A case in point is the much-publicized neutron bomb, which has been promoted as a specialized antitank weapon, since neutrons can penetrate tank armor and kill the crew. It is frequently overlooked that the neutron bomb is, in fact, a nuclear weapon with significant yield. While it does emit some ten times as many neutrons as a comparable "ordinary" small nuclear weapon, it also kills by blast, heat, and prompt radiation. For instance, one of the proposed neutron warheads for the Lance missile has a yield of 1 kiloton, which would produce the same levels of blast damage experienced at Hiroshima at a little less than one-half the distance from the point of detonation.

An attack on tanks near a populated area, or a targeting error in the heat of battle, would clearly have a far-reaching effect on civilians and structures in the vicinity. Moreover, the lethal effects of the neutrons are not sharply defined. There would be attenuation by intervening structures or earth prominences, and there is a wide gap (from 500 to 10,000 rems) between a dose that would eventually be fatal and one that would immediately prevent a soldier from continuing combat. Under actual war conditions, no local commander, much less a national decision-maker, could readily tell whether a neutron weapon or some other kind of nuclear weapon had been employed by the enemy. Thus, the threat of escalation from local to all-out conflict, the problems of collateral damage of nuclear weapons, and the disastrous consequences of errors in targeting are not changed by the nature of the nuclear weapons.

In short, whatever the utility of the neutron bomb or any other "tactical" nuclear weapon in *deterring* Soviet conventional or nuclear attack, any actual use of such weapons is extremely unlikely to remain limited. We come back to the fundamental point that the only meaningful "firebreak" in modern warfare, be it strategic or tactical, is between nuclear and conventional weapons, not between unilaterally proclaimed categories of nuclear weapons.

The thesis that we live in an inherently MAD world rests ultimately on the technical conclusion that effective protection of the population against large-scale nuclear attack is not possible. This pessimistic technical assessment, which follows inexorably from the devastating power of nuclear weapons, is dramatically illustrated by the fundamental difference between air defenses against conventional and nuclear attack. Against bombers carrying conventional bombs, an air-defense system destroying only 10 percent of the incoming bombers per sortie would, as a practical matter, defeat sustained air raids, such as the ones during World War II. After ten attacks against such a defense, the bomber force would be reduced to less than one-third of its initial size, a very high price to pay, given the limited damage from conventional weapons, even when more than 90 percent of the bombers penetrate. In contrast, against a bomber attack with nuclear bombs, an air defense capable of destroying even 90 percent of the incoming bombers of each sortie would be totally inadequate, since the damage produced by the penetrating 10 percent of the bombers would be devastating against urban targets.

When one extends this air-defense analogy to ballistic-missile defenses intended to protect population and industry against large numbers of nuclear missiles, it becomes clear that such a defense would have to be almost leakproof, since the penetration of even a single warhead would cause great destruction to a soft target. In fact, such a ballistic-missile defense would have to be not only almost leakproof but also nationwide in coverage, since the attacker could always choose the center of population or industry he wished to target. The attacker has the further advantage that he cannot only choose his targets but also decide what fraction of his total resources to expend against any particular target. Thus, an effective defense would have to be extremely heavy across the entire defended territory, not at just a few priority targets. The technical problem of providing an almost leakproof missile defense is further compounded by the many technical measures the attacking force can employ to interfere with the defense by blinding or confusing its radars or other sensors and overwhelming the system's traffic-handling capacity with decoys.

When these general arguments are reduced to specific analysis, the conclusion is inescapable that effective protection of the population or industry of either of the superpowers against missile attack by the other is unattainable with present ABM (antiballistic missile) defense technology, since even the most elaborate systems could be penetrated by the other side at far less cost. This conclusion is not altered by prospective improvements in the components of present systems or by the introduction of new concepts, such as lasers or particle beams, into system design.

These conclusions, which address the inability of ballistic-missile defense to eliminate the MAD character of the strategic relationship, do not necessarily apply to defense of very hard point targets, such as missile silos or shelters for mobile missiles. The defense of these hardened military targets does offer a more attractive technical opportunity, since only the immediate vicinity of the hardened site needs to be defended, and the survival of only a fraction of the defended silos is necessary to serve as a deterrent. Thus, the technical requirements for the system are much less stringent than for population or industrial defense, and a much higher leakage rate can be tolerated. When these general remarks are translated into specific analysis that takes into account the many options available to the offense, hard-site defense still does not look particularly attractive. Moreover, such a defense, even if partially successful, would not prevent the serious collateral fallout effects from the attack. Nevertheless, the fact that these systems are technically feasible, and are advocated by some as effective, tends to confuse the public on the broader issue of the feasibility of urban defense against ballistic missiles.

The United States has a substantial research-and-development effort on ballistic-missile defenses of land-based ICBMs as a possible approach to increasing the survivability of this leg of the strategic triad. The only program under serious consideration that could be deployed in this decade is the so-called LOAD (Low Altitude Defense) system. This system, which would utilize very small hardened radars and small missiles with small nuclear warheads, is designed to intercept, at very close range, those attacking missiles that might detonate close enough to the defended ICBM to destroy it. This last-ditch defense is possible with nuclear weapons, since the defended target is extremely hard and can tolerate nuclear detonations if they are not too close. While such a system for the defense of hard sites is technically feasible, there has been serious question as to whether it would be cost-effective in defending the MX in fixed Titan or Minuteman silos, since the system could be overwhelmed relatively easily. In the case of the defense of a mobile MX in a multiple-shelter system, the economics of the exchange ratios is substantially improved if the location of the mobile MX and mobile defense system are in fact unknown to the attacker; however, there are serious questions whether the presence of radiating radar systems might not actually compromise the location of the MX during an attack.

Looking further into the future, the U.S. research program is considering a much more sophisticated "layered" system for hard-site defense. The outer layer would consist of an extremely complex system using infrared sensors that would be launched on warning of a Soviet attack to identify and track incoming warheads. Given this attack information, many interceptors, each carrying multiple, infrared-homing rockets with nonnuclear warheads, would be launched against the cloud of incoming warheads and would attack them well outside the atmosphere. The warheads that leaked through this outer exoatmospheric layer would then be engaged by a close-in layer along the lines of the LOAD last-ditch system described above.

It has been suggested that the outer layer exoatmospheric system might evolve into an effective area defense for population and industry. Actually, there are many rather fundamental technical questions that will take some time to answer about the ability of such a system to work at all against a determined adversary in the time frame needed to deploy it. For example, such a system would probably be defeated by properly designed decoys or blinded by nuclear explosions and, above all, may well be far too complex for even prospective control capabilities to operate. Whatever the value of these types of systems for hard-site defense to support the MAD role of the deterrent, it is clear that the system holds no promise for the defense of population or industry and simply illustrates the technical difficulty of dealing with that problem.

While the government struggles with the much less demanding problem of whether it is possible to design a plausible, cost-effective defense of hardened ICBM silos, the public is bombarded with recurring reports that some new technological "breakthrough" will suddenly generate an "impenetrable umbrella" that would obviate the MAD strategic relationship. Such irresponsible reports usually rehash claims for "directed-energy" weapons that are based on the propagation of extremely energetic beams of either light (lasers) or atomic particles propagated at the speed of light to the target. Some of the proposals are technically infeasible, but, in all cases, one must remember that, for urban defense, only a system with country-wide coverage and extraordinarily effective performance would have an impact on the MAD condition. To constitute a ballistic-missile defense system, directed-energy devices would have to be integrated with detection and tracking devices for the incoming warheads, an extremely effective and fast data-handling system, and the necessary power supplies for the extraordinarily high demand of energy to feed the directed-energy weapons, and they would have to be very precisely oriented to score a direct hit to destroy the target as opposed to nuclear warheads, which would only have to get in the general vicinity in order to destroy the target.

There are fundamental considerations that severely limit the application of directed-energy weapons to ballistic-missile defense. Particle beams do not penetrate the atmosphere. Thus, if such a system were ground-based, it would have to bore a hole through the atmosphere, and then the beam would have to be focused through that hole in a subsequent pulse. All analyses have indicated that it is physically impossible to accomplish this feat stably. Among other things, laser systems suffer from the fact that they can only operate in good weather, since clouds interfere with the beam.

These problems caused by atmospheric effects could be avoided by basing the system in space. Moreover, a space-based system has the desirable feature of being able potentially to attack missiles during the vulnerable launch phase before the reentry vehicles are dispersed. However, space-based systems require that a very complex system with a large power requirement be placed into orbit. Analysis indicates that a comprehensive defensive system of this type would require more than a hundred satellites, which, in turn, would require literally thousands of space-shuttle sorties to assemble. It has been estimated that such a system would cost several hundred billion dollars. Even if the control mechanisms were available to operate such a system, there are serious questions as to the vulnerability of the satellites to physical attack and to various measures that would interfere with the system's operation. In short, no responsible analysis has indicated that, for at least the next two decades, such "death-ray weapons" have any bearing on the ABM problem, or that there is any prospect that they would subsequently change the MAD character of our world.

Defense against aircraft further illustrates the inherently MAD nature of today's world. Although the Soviets have made enormous investments in air defense, the airborne component of the U.S. strategic triad has not had its damage potential substantially reduced. Most analyses indicate that a large fraction of the "aging" B-52 fleet would penetrate present Soviet defenses, with the aid of electronic countermeasures and defense suppression by missiles. It is true that the ability of B-52s to penetrate will gradually be impaired as the Soviets deploy "look-down" radar planes similar to the much publicized AWACS (Airborne Warning and Control System). However, these systems will not be effective against the air-launched cruise missiles whose deployment on B-52s will begin shortly; their ability to penetrate will not be endangered until a totally new generation of Soviet air defenses enters the picture. At that time, one can foresee major improvements in the ability of bombers and cruise missiles to penetrate through a number of techniques—in particular, the so-called "stealth" technology, which will reduce by a large factor the visibility of airplanes and cruise missiles to radar.

In short, there is little question that, in the defense–offense race between air defenses and the airborne leg of the triad, the offense will retain its enormous damage potential. For its part, the United States does not now have a significant air defense, and the limited buildup proposed in President Reagan's program would have little effect on the ability of the Soviets to deliver nuclear weapons by aircraft against this country. Consequently, the "mutual hostage" relationship between the two countries will continue, even if only the airborne component of the triad is considered.

It is sometimes asserted that civil defense could provide an escape from the consequences of the MAD world and make even a general nuclear war between the superpowers winnable. This assertion is coupled with a continuing controversy about the actual effectiveness of civil defense and the scope of the present Soviet civil-defense program. Much of this debate reflects the complete failure of some civil-defense advocates to comprehend the actual consequences of nuclear war. There is no question that civil defense could save lives and that the Soviet effort in this field is substantially greater than that of the United States. Yet, all analyses have made it abundantly clear that, to have a significant impact in a general nuclear war, civil defense would require a much greater effort than now practiced on either side, and that no amount of effort would protect a large portion of the population or the ability of either nation to continue as a functioning society.

There is evidence that the Soviets have carried out a shelter program that could provide fallout protection and some blast protection for about 10 percent of the urban population. The only way even to attempt to protect the bulk of the population would be complete evacuation of the entire urban population to the countryside. Although to our knowledge there has never been an actual urban evacuation exercise in the Soviet Union, true believers in the effectiveness of Soviet civil defense point to the alleged existence of detailed evacuation plans for all Soviet cities. Yet, when examined in detail, there are major questions about the practicality of such evacuation plans.

The U.S. Arms Control and Disarmament Agency has calculated, using a reasonable model and assuming normal targeting practices, that in a general war there would still be at least 25 million Soviet fatalities, even with the general evacuation of all citizens and full use of shelters. Such estimates obviously depend on the model chosen: some have been lower, but others, by the Defense Department, have been considerably higher. The time required for such an all-out evacuation would be at least a week. This action would guarantee unambiguous strategic warning and provide ample time for the other side to generate its strategic forces to full alert, which would result in a substantially greater retaliatory strike than would be expected from normal day-to-day alert. If the retaliatory strike were ground-burst to maximize fallout, fatalities could rise to 40 to 50 million; and if part of the reserve of nuclear weapons were targeted against the evacuated population, some 70 to 85 million could be killed. Until recently, little has been said about the hopeless fate of the vast number of fallout casualties in the absence of organized medical care or about what would become of the survivors with the almost complete destruction of the economic base and urban housing.

Finally, there is no evidence that the Soviets are carrying out industrial hardening or are decentralizing their industry, which remains more centralized than U.S. industry. This is not surprising, since there is nothing they can do that would materially change the inherent vulnerability of urban society in a MAD world.

In sum, we are fated to live in a MAD world. This is inherent in the tremendous power of nuclear weapons, the size of nuclear stockpiles, the collateral damage associated with the use of nuclear weapons against military targets, the technical limitations on strategic area defense, and the uncertainties inherent in efforts to control the escalation of nuclear war. There is no reason to believe that this situation will change for the foreseeable future, since the problem is far too profound and the pace of technical military development is far too slow to overcome the fundamental technical considerations that underlie the mutual-hostage relationship of the superpowers.

What is clear above all is that the profusion of proposed NUTS approaches has not offered an escape from the MAD world but, rather, constitutes a major danger in encouraging the illusion that limited or controlled nuclear war can be waged free from the grim realities of a MAD world. The principal hope at this time will not be found in seeking NUTS doctrines that ignore the MAD realities but, rather, in recognizing the nuclear world for what it is and seeking to make it more stable and less dangerous.

Misperceptions about Arms Control

Those assembled here are working on pure and applied science. Applied work relates both to civilian and to military problems. With respect to the latter, there has been a great deal of controversy about what is wise or unwise, reasonable or unreasonable to pursue. I would like to remind you of the remark reputedly made at the turn of the century by the great mathematician David Hilbert: "Nobody can prevent mathematics from being applied."

I will go even further by stating that few would disagree that research, pure and applied, is necessary and desirable. Ignorance of fundamental principles and facts never helps. As scientists, but also as citizens, we must assume that, however fallible human judgment may be, we make better decisions, on the average, if we know what we are talking about than if we don't. Yet most would also agree that, when technical work, including such work applied to military purposes, goes beyond the acquisition of fundamental knowledge and leads to development and deployment, then some restraints are needed to prevent technology from dominating mankind. Society should decide how far applications of technology should go.

In the military area, this directly points to the need for nations to work together to agree on the means for managing an otherwise unrestrained arms competition. As the quantity and quality of military hardware in-

Banquet address given at the Institute of Electrical and Electronics Engineers at the Particle Accelerations Conference, March 18, 1987.

crease, the security of all parties may, in fact, be impaired. The mission of the discipline of arms control is to prevent this latter outcome.

Here I am talking about the many misperceptions that are floating around concerning this subject. My purpose is to slay some of these dragons, which have been impediments to progress. Let me list some of them:

1. The Soviets have been cheating left and right on past arms-control agreements, so what is the point in negotiating new ones?
2. There has been no significant progress in arms control in the past, so we might as well give it up.
3. The traditional interpretation of the ABM Treaty effectively kills the Strategic Defense Initiative and therefore prevents the accumulation of the knowledge necessary for making intelligent future decisions on what should be done about defense against ballistic missiles.
4. Whenever we negotiate with the Russians, we get "outnegotiated," so that we lose and they gain; the Russians are better chess players than we are.
5. Progress in arms control will give us a false sense of security.

As for the first dragon, the reason why the perception that the Soviets are cheating left and right is widespread is that the Reagan administration has been engaged in a systematic search for Soviet violations. This, in turn, has induced Congress to ask the executive branch for an annual report on such violations. Such reports have become one-sided "briefs of the prosecution" rather than balanced analyses. This "violations hunt" has taken the place of a reasoned assessment of the degree of compliance by all parties, including the Americans as well as the Soviets, to past arms-control agreements. Had such a net assessment of compliance been conducted and widely publicized, then we would have come to the important realization that compliance with past arms-control agreements has been amazingly good.

Let me draw an analogy from civilian life: We have lots of laws on the books dealing with environmental pollution; the Congress and other legislative bodies have to examine continuously the wisdom of past laws and future provisions. If all that was done was to make an inventory of violations of existing environmental law, we would get a very deceptive picture indeed. The dominant question is not how many people disconnect their emission controls or how many industries dump waste illegally into

Lake Michigan; the overriding issue is, first, whether the pollution picture would be worse without regulation than with regulation, and, second— and more important—whether we are gaining or losing in our battle against environmental pollution. So it is with arms control. In addition to examining the violations record of all parties, we should also judge whether the arms race would be even worse in the absence of existing agreements, and, more important, whether there is any expectation that an arms-control regime has any hope of reversing the escalatory spiral of arms competition.

Thus, just pinpointing violations—real, conjectured, or manufactured—and ignoring the total compliance record is destructive. But there are limits to the analogy between arms control among sovereign nations and domestic law. When violations do occur at home, the violators can be haled into court. If legislation is ambiguous, then the courts can intepret the law. But compliance with international arms-control agreements among sovereign states at some level must remain voluntary. Therefore, in the present international order, one must establish mechanisms for airing mutual complaints and for resolving ambiguities. If there is a will on all sides to make an arms-control agreement work, then such constructive mechanisms can work. But such a will can only be created if all sides of an agreement believe that their national security is improved through the arms-control agreement in question—arms control is not a zero-sum game, in which one side loses and the other side gains. Unless all sides believe they gain through limiting the arms race, there is little hope.

Demonstrable violations can, of course, not be condoned under any circumstances, just as running a red light when nobody is coming is an unacceptable violation. Violations breed disrespect of treaties or laws, as the case may be. However, most alleged violations are really areas of dispute in which arms-control provisions have been inadvertently or deliberately left ambiguous, and where gray areas of conduct exist. Most of the alleged violations about which you read in the government's compliance report are in the ambiguous or disputed area. Only one item cited in the report—the Krasnoyarsk Radar—is a demonstrable violation. Among the disputed issues is conduct by both the Soviets and the United States in relation to SALT II and the ABM Treaty, as well as other arms-control agreements. Until the recent conversion by the United States of an additional B-52 bomber carrying cruise missiles, there was in fact no demonstrable violation at all by either the Soviets or the United States of the unratified SALT II agreements; therefore, the decision by the United States to discontinue compliance with SALT II on the basis of alleged Soviet noncompliance lacks a factual basis. There should be no doubt that both the Soviet Union and the United States have decommissioned many military systems to stay within SALT II limits or that projections of future weaponry reach much larger figures if SALT II were to continue to lapse than if it were to continue to remain a commitment. The United States has failed to cooperate with the problem-solving machinery set up as part of the SALT process to resolve disputes in interpretation or to challenge action of the other party.

There is *no* factual basis for charging, as the Reagan administration has done, that the Soviets are "likely to have violated" the Threshold Test Ban Treaty. In fact, such charges are the result of confusing unavoidable uncertainties in measurement with "likely violations." There is *no* factual basis for charging, as the Reagan administration did last week, that "the USSR may be preparing an ABM system for the defense of its national territory," as prohibited by the ABM Treaty.

Whatever one's views about who is right and who is wrong in the interpretation of disputed clauses of past arms-control agreements, there is no question that the alleged violations are of minor military importance in the overall context of the vast military might of all sides. Thus, the overemphasis on violations out of context with the overall compliance picture is a deliberate political move, not a matter of military necessity. This public "witch hunt" for violations has improperly undermined confidence in the arms-control process.

As for the second dragon—yes, Virginia, there *have* been important arms-control agreements. Arms control has not been a failure, although, indeed, past agreements in the aggregate have made only a small dent in the U.S.–Soviet competition in arms. Twenty arms-control agreements between the United States and the Soviet Union exist today. Of these, the two having the most profound impact are the Non-Proliferation Treaty and the ABM Treaty.

Almost any knowledgeable person who had been asked, in the 1960s, how many nations would have exploded nuclear weapons by 1987 would have answered, "About twenty," and yet the correct answer history has shown us, is six. This may be small comfort, but it must be realized that no technical barriers can be erected against the acquisition of nuclear weapons; ultimately, nations must be persuaded that they are safer without them than with them. This is a political problem—the Non-Proliferation Treaty has been of great help in reducing candidates for admission into the Nuclear Club—but progress by the Superpowers in limiting their own nuclear arsenals will, in the long run, have the greatest leverage in limiting the spread of nuclear weapons.

Then there is the beleaguered ABM Treaty, which was signed about fifteen years ago. It has served the nation's security extremely well. It has prevented a destructive and costly offense–defense race in ballistic missiles; yet it has permitted research on both sides to make sure that the technical assumptions under which the treaty was negotiated are still valid today. Contrary to popular belief, nothing has turned up to change these basic assumptions. Contrary to frequent misstatements implying that ABM research started only with President Reagan's star wars initiative, such research has been in process as long as ballistic missiles have been around. The treaty has made it possible for the allies of the United States to have confidence that their own independent forces could reach their targets; thus, the dependence of these allies on U.S. nuclear intervention has been decreased. Both sides have deployed early-warning radars around their perimeters, but neither side has now deployed—and could not deploy, under the treaty—ballistic-missile defenses to blunt significantly the deterrent missiles of the other side.   Without the ABM Treaty, the nuclear arms race would have been a great deal worse, but I agree that, with 60,000 warheads worldwide, it is bad enough.

And now to the third dragon: The ABM Treaty has been interpreted, for all these years, as prohibiting the testing, development, and deployment of mobile and space-based defenses. Now we are facing what was first called a "reinterpretation," or what the Reagan administration has euphemistically called the "legally correct interpretation," changing all this. This, in turn, has raised another misconception, or at least a question: Does the ABM Treaty, as traditionally interpreted, "kill" SDI, and does it prevent research to examine the promise of defense against ballistic missiles? My answer is "No."

The ABM Treaty, as traditionally interpreted, permits three classes of research under the arms-control process:

1. any research in the laboratory;
2. any research and development using fixed means of interception at specified test sites (which, for the United States, are Kwajalein and White Sands Missile Range); and
3. any research on more exotic systems, provided it does not cross

the boundary leading to actual development or test of components of systems to be deployed.

This latter category, obviously, is subject to some gray-area interpretation as to what a component of an operational system is, but I believe you will agree that these three categories permit a huge amount of work. Last year [1986], General Abrahamson said publicly that the traditional interpretation would not inhibit any SDI work before the early 1990s. The only reason to go beyond these limits would be to go to deployment of systems that are not technically ready, or to deliberately make demonstrations whose targets are not enemy missiles but are the ABM Treaty itself. Thus, the claim so frequently heard to the effect that we need either to abrogate the ABM Treaty or to come up with a new, broader interpretation that will permit us to go to full development and testing of spaceborne systems cannot be justified as being necessary to acquire much needed knowledge for future decisions. Stating that the ABM Treaty in its traditional interpretation "kills the SDI" is simply not true.

Now let us slay the fourth dragon: Whenever we meet with the Soviets, we are "outnegotiated" because the Soviets are so much better at negotiation than we are. This is just plain nonsense. Arms-control negotiators are a short leash. They receive instructions from either Washington or Moscow that do not permit much latitude. If we are outnegotiated, it means that there is a lack of wisdom in our deliberations in Washington. The negotiating process itself is not a chess game, in which each delegation can calculate as many moves ahead as they wish and freely act accordingly. In a successful arms-control agreement, both sides win. We do not have one winner and one loser, or a "draw" as is the case in chess. If this were really the problem, maybe what we need is a "Strategic Chess Initiative" to correct the matter.

The fifth and final dragon I would like to attack is the statement that arms control is dangerous because it will lull us into a false sense of security and therefore cause us to provide inadequately for our defense. I hope indeed that this argument is false. It drastically attacks our form of government in that it states that, if we do the right thing by damping down the clearly destructive, dangerous, and wasteful arms competition, then our decision-making processes will be unable to deal intelligently with our national security. I can think of no graver disservice to the democratic process than to claim that it cannot function unless it is artificially stimulated by such a destructive competition.

There is no conflict between research, military or otherwise, which replenishes our pool of knowledge and which enables us to face intelligently the important decisions affecting the future of this country and the world, and effective arms control, which limits the dangers and burdens of armament. Stable agreements can be reached by negotiation with adversaries with whom we have conflicting interests, in many parts of the world. All countries have a common interest that overrides their conflicting interest, and that is survival.

The process of reaching such agreements must be based on realistic assessments of what technology can do. In fact, the biggest problem besetting the process of formulating military policy today is the ever-widening gap between the political perception of what technology is capable of doing and the technical realities. The types of distortions of reality that I have summarized, as well as the misconceptions about what science can accomplish in solving essentially political problems, serve no one.

I believe that the largest public responsibility, irrespective of whatever convictions individual scientists may have about the proper course of action, is to close this gap between perception and reality. Policy, particularly in the field of national security, should be based on a reasonable assessment of science and technology. Political authority can no more decree, as was attempted under Stalin and Khrushchev during the heyday of Lysenko, that the established rules of genetics be abrogated in favor of a crackpot, but politically favored, view of the science, than a president can decree that cancer be abolished without understanding the biological rules underlying the disease. National security is not served if a president simply ordains that scientists make nuclear weapons impotent and obsolete without knowing of a scientifically valid way for them to do so. In the current international order, true national security needs military research—but not misguided policy or oratory based on anticipating results of such work that have not yet been obtained and that are very unlikely ever to be obtained. True national security needs realistic arms control, not misperceptions about arms control.

Arms Control, Compliance, and the Law

WITH GEORGE BUNN

The United States and Soviet Union have signed the INF treaty. Arms-control talks are in various stages. The bilateral START forum and multilateral negotiations on chemical weapons continue in Geneva. Discussions about banning nuclear tests are resuming for consideration of means to tighten verification, as demanded by the Reagan administration as a condition for ratification of the Threshold Test Ban Treaty and the Peaceful Nuclear Explosions Treaty, both signed over a decade ago. Many observers hope that stricter limits on nuclear testing will also result. Partly in consequence of the INF Treaty and the demonstrated willingness of the Soviet Union to accept asymmetric reductions, pressures for decreasing the dangers and burdens of conventional weapons in Europe will almost certainly lead to further arms-control negotiations toward that end.

The expected degree of compliance with agreements or treaties (we use the two terms interchangeably) under negotiation must be inferred from the record of the past as well as from the expectations that are raised by the willingness of the parties to dedicate increased resources to verification, to accept inspections that are more intrusive, to exhibit greater openness, and to provide for cooperative verification measures. Yet such

A working paper of the Center for International Security and Arms Control, Stanford University, May 1988.

expectations for compliance must be tempered by the recognition of certain fundamental factors, which we will discuss here.

The public discourse dealing both with the record of the past and with the expectation of future compliance has too often been simplistic. Standards are invoked for compliance with arms control that one would not dream of applying to compliance with domestic law. Overall compliance with recent arms-control agreements has, in fact, been good, although there are numerous disputes on specific items of conduct. Yet the perception of past compliance has been distorted by public charges of violation that are either downright mistaken or taken out of context or that stem from ambiguous treaty language or conduct.

In our discussion of the compliance process, we invoke a broad standard that we believe should apply: An arms-control agreement is worthy of support by the United States if, with adequate verification, its result improves U.S. national security. Under this standard, we analyze the compliance process relating to arms control, and we examine its similarities and dissimilarities, with respect to the domestic statutory process.

In comparing arms control with domestic law, we are making the conscious assumption that a world with unconstrained nuclear-arms competition is no more tolerable than a domestic society without law. Arms-control agreements attempt to manage the weapons activities among nations so as to increase each nation's security and decrease the danger of devastating war and the burden of armaments. As a middle course between unilateral disarmament and an unfettered arms race, arms control is a complex and difficult process. But so is domestic law, as a middle course between individual license and arbitrary state power. Yet, in a real sense, both arms control and domestic law are essential elements of a durable civilization. Thus, a comparative evaluation should help to clarify the balance of values inherent in arms control. The choice cannot be between perfect compliance with arms control and no arms control at all. Such a choice is no more feasible than one between perfect law enforcement and an absence of law.

We survey the record of compliance with some past arms-control agreements and come to the conclusion, perhaps surprising to some, that recent arms-control agreements have, in fact, effectively governed the conduct of the Soviet Union and the United States. We also conclude that, although arms-control violations should in no way be condoned, verification and compliance problems are frequently exaggerated. In fact, most of the issues that arise stem from dissatisfaction with the basic strictures of the agreements; allegations of noncompliance often serve as a cover for such dissatisfaction.

The Arms-Control Process

The question of compliance is interwoven with the entire process of arms control. We consider compliance as it relates to the three major phases of the arms-control process: (1) negotiated drafting of the provisions of a treaty with due regard to future compliance; (2) designing of verification measures to ascertain the facts about the degree of compliance with a treaty; and (3) implementation of an agreement in a manner to maximize compliance.

Drafting an Agreement. The drafting of a basic agreement must confront conflicting values that are related to future compliance. First, there is the matter of "precision versus breadth." If the provisions of the treaty are drawn too precisely in terms of the technical military situation at the time of drafting, ambiguities and loopholes are apt to develop as that situation changes. If the treaty is drawn too broadly, compliance will face gray areas whose management should have been anticipated in the treaty.

Second, there is the matter of "creative ambiguity." When there are real differences in position among the parties during the treaty-drafting process, it may be expedient, or even unavoidable, to bridge these differences by drafting ambiguous terms rather than reconciling the difference. However, such ambiguity invites future problems.

An example of deliberate treaty ambiguity concerns encryption and the SALT II Treaty. During negotiations and drafting, both the United States and Soviet Union recognized that interception of telemetry data during missile testing is a valuable asset for verifying the characteristics of missile systems. They also recognized that encrypting such data can, at least in part, deny such information and that other methods, such as encapsulating and then dropping a data container for later recovery or minimizing the use of telemetry during the testing process, can inhibit foreign acquisition of test data. Agreement on total prohibition of telemetry encryption proved impossible (but has become possible for START by the agreement reached in the December 1987 summit between President Reagan and General Secretary Gorbachev). Instead, the two sides adopted a compromise, that telemetry may not be denied "whenever such denial impedes verification of compliance with the provisions of the Treaty" (SALT II, Article XV, Second Common Understanding). The Soviets subsequently greatly increased the use of encryption of telemetry. When the United States challenged this practice, the Soviets offered to open up those channels that the United States deemed essential to verifying treaty compliance. The United States was unwilling to identify the specific channels, arguing that such identification would compromise "sources and methods" of intelligence collection. In this case, the United States made a deliberate choice to retain the ambiguous compromise as preferable to either an outright ban on encryption or no ban at all.

Third, there is the matter of compliance itself. The substantive provisions of a treaty should facilitate compliance. Drafting them may thus require a balance of values between the most desirable weapons restraints and the optimum verification procedures. For instance, the "double-zero" provisions of the INF agreement offer substantial verification advantages over numerical limits other than zero. Because burdensome identification and precise counting of permitted missiles are not required, the observation of any single intermediate-range ballistic missile would constitute proof of a clear violation.

Fourth, there is the matter of expectations of change in technology. For example, the 1972 ABM Treaty directly anticipates possible future ABM systems based on "other physical principles." At the time the treaty was drafted, an ABM system consisted of early-warning radars, acquisition and battle-management radars, interceptor missiles and nuclear warheads and their launchers, and a data-management control system tying all the subsystems together. The drafters recognized that other technologies might show promise in the future. In fact, several spaceborne ABM systems had been studied but found wanting, and research on new ABM technologies was in process. Article II of the treaty acknowledges *both* the then-existing and the expected future technologies by defining an ABM system as "a system to counter strategic missiles or their elements in flight trajectory, currently consisting" of radars and interceptors and their launchers. Further, Article V and the annexed Agreed Statement D recognized the possible impact of future technologies on the treaty.

Article V explicitly forbids development, testing, or deployment of "ABM systems or components which are sea-based, air-based, space-based, or mobile land-based." This statement is generic, not tied to a specific technology. At the time of drafting, no systems built or planned fitted into these categories. The treaty as amended in 1974 prohibits deployment of ABM systems or components except for limited placement by each party at a single *fixed* site on land. In order to forestall the deployment of systems and components using "other physical principles" at this site, Agreed Statement D requires U.S.–Soviet discussion in case such exotic systems are "created" (meaning, by consensus, "developed and tested") as a result of the permitted research-and-development activities on fixed, land-based ABM systems.

Notwithstanding the treaty text, specifically designed to anticipate technological progress, the Reagan administration announced on October 6, 1985, its "reinterpretation" of the ABM Treaty, maintaining that testing and development involving new physical concepts are permitted by the treaty and that only deployment is foreclosed. We do not deal with the resulting dispute in this paper.[1]

Even within the traditional interpretation of the ABM Treaty, a considerable gray area surrounds what might reasonably be considered allowed or forbidden conduct; such treaty terms as "ABM component," "testing in an ABM mode," and "provide a base for" permit some latitude of interpretation. The treaty incorporated a consultative mechanism, the Standing Consultative Commission (SCC), to deal with such gray areas.

The Strategic Defense Initiative Organization (SDIO) has submitted plans for tests that it states comply with the traditional interpretation of the ABM Treaty; in accordance with congressional action, SDI test activities for fiscal year 1988 will be so restricted. A paradoxical situation results. Some tests are of the nature of demonstrations; their purpose is to persuade the public that feasible ABM technologies are close at hand. Yet, according to the traditional interpretation of the treaty, testing or development of an ABM component based on other physical principles—a component having the function of one in existence at the time the treaty was negotiated—would constitute a violation of the treaty. Thus, on the one hand, the administration claims limited effectiveness of the projected

---

[1] The two main protagonists, Abraham Sofaer, legal adviser to the U.S. State Department, and Sam Nunn, chairman of the Senate Armed Services Committee, summarized the debate between the Senate and the executive branch (*Washington Quarterly* 10, no. 4 (1987): 45, 59). See also *The Anti-Ballistic Missile Treaty Interpretation Dispute* (Association of the Bar of the City of New York, January 1988) and R.L. Garthoff, Policy Versus the Law: *The Reinterpretation of the ABM Treaty* (Washington: Brookings Institution, 1987); these references cite documents from the negotiations and longer legal analyses from the executive and legislative branches.

tests in order to show compliance with the ABM Treaty, whereas, on the other hand, it claims that these very tests advertise technological readiness. This ambivalence produces conflicts among official statements and programs and induces doubts at home and abroad that particular tests comply with the treaty. The situation calls for invoking mechanisms for consultation with the Soviets in order to reach agreement on the issue of allowed versus forbidden conduct. Thus far, the administration has not been willing to invoke these mechanisms, but the Shevardnadze–Shultz discussions at the end of March 1988 indicate that the administration acknowledges that the issue cannot be separated from a START agreement.

Such examples show how conflicts of value arise during the drafting of treaties and how subsequent compliance disputes become unavoidable. Yet, in these examples, the provisions containing potential for disputes were considered preferable to no terms at all or to terms stating precise but possibly perishable boundaries between allowed and forbidden conduct.

Designing Verification Measures. The second phase of the arms-control process, verification, has received so much prominence in public discourse that many consider it to be the only essential element governing compliance. "Trust and verify," to quote President Reagan at the 1987 summit, is hardly a prescription for compliance. The goals of verification are three: (1) detection of a class of "suspicious events," that is, possible violations; (2) identification of these events to determine whether or not they are prohibited; and (3) measurement to verify compliance or noncompliance with a treaty's quantitative limits.

The tools of verification are the collection and analysis of information needed to monitor compliance with arms-control provisions. The means to collect such information, which analysts have discussed extensively, include (1) national technical means (NTM), such as satellite surveillance, radar surveillance from locations outside the boundaries of the country monitored, air sampling, teleseismic geophysical observations, and interception of communications and other signals; (2) cooperative means of verification, such as deliberate openness of certain features of military systems, specific channeling of military products through agreed-upon check points, and noninterference with means of verification; (3) technical devices emplaced on the territories of parties to the agreement; (4) on-site inspection; and (5) "soft" methods of verification, such as agents, interviews with emigres, and information leaks.[2]

Such means can rarely be specifically dedicated to verification of arms control; there is a necessary overlap between intelligence activities and arms-control verification. The general collection of intelligence benefits verification of arms control, and arms-control verification spins off information about foreign activities not related to arms control. Indeed, during negotiation, one side may denounce the other's position on verification as attempted espionage.

Specification of verification measures in an arms-control treaty requires a balance of several values. First, there is the matter of intrusion. Intrusiveness of verification measures can conflict with privacy and protection of military and commercial secrets. Historically, the Soviet Union has erected much higher barriers against intrusion for the sake of verification than has the United States. This situation, however, has changed substantially in the Gorbachev era. Under *glasnost*, the Soviets appear to be willing to accept, at times, more inspection and other intrusions than the Western powers.

Second, there is the matter of cost. Verification provisions can require large resources of material and manpower. At some point, the question must be faced whether national security is better served if such resources are dedicated to verification of arms-control agreements or to some other goal.

Third, the drafting and negotiation of verification provisions require an understanding of the *standards* that verification will have to meet. Little is served by simplistic statements that a proposed agreement can or cannot be verified. In 1977, the Congress established "adequate verification" as a legislative standard for arms-control agreements.[3]

The House had earlier adopted the term "effective," but the House and Senate agreed to "adequate" in conference, and the statute so reads. "Adequate" is, of course, a subjective term. In explaining the standard, Senator Cranston, the floor manager of the amendment, said, "In deter-

---

[2] See, for example, R.A. Scribner, T.J. Ralston, and W.D. Metz, *The Verification Challenge: Problems and Promise of Strategic Nuclear Arms Control Verification*, project of the Committee on Science, Arms Control, and National Security of the American Association for Advancement of Science in cooperation with the Center for International Security and Arms Control, Stanford University (Boston: Birkhäuser, 1985).

[3] See *United States Statutes at Large* 91 (August 17, 1977): 871. The term "adequate verification" is now in Section 37(a) of the Arms Control and Disarmament Act, *United States Code* 22 (1982 edition): 2577.

mining the adequacy of verification, a *balance* will have to be cast...between the degree of risk posed by possible violation of an agreement and the gains to U.S. security flowing from the restraints imposed by the agreement on other countries."[4]

Secretary of Defense Harold Brown gave the Carter administration's interpretation of "adequate verification" in his testimony during the SALT II hearings: "The relevant test is not an abstract ideal, but the practical standard of whether we can determine compliance *adequately to safeguard our security*—that is, whether we can identify attempted evasion if it occurs on a large enough scale to pose a significant risk, and whether we can do so in time to mount a sufficient response. Meeting this test is what I mean by the term "adequate verification."[5]

On the basis of the compliance history since Brown's testimony, we would add that verification, to be adequate, should catch not just violations that threaten national security but also those that, although not so threatening, nevertheless undermine an important purpose of a treaty.

In 1981, the Reagan administration changed the standard from "adequate" to "effective," signaling that it wished to increase verification requirements. In its arms-control impacts statements for fiscal year 1987, the administration defined effective means of verification as those that "enable the United States to know with confidence whether the terms of agreements are being honored."[6] Note that consideration of military importance, national security, or the purposes of a treaty are not included in this standard; we consider the omission unfortunate, although it was largely corrected in administration testimony to the Senate in support of the INF Treaty. Paul Nitze, special adviser to the president on arms control, defined "effective" as meaning, "If the other side moves beyond the limits of the treaty in *any militarily significant way*, we would be able to detect such violation in time to respond effectively, and thereby deny the other side the benefit of the violation."[7]

Admiral William Crowe, chairman of the Joint Chiefs of Staff, actually used "adequate" in testimony before the Senate Committee on Armed Serv-

---

[4] *Congressional Record* 123 (1977): 19515 (emphasis added).

[5] U.S. Senate, Foreign Relations Committee, *Hearings on the SALT II Treaty*, 96th Cong., 1st sess. (1979), part 2, 241 (emphasis added).

[6] U.S. President, *Fiscal Year 1987 Arms Control Impact Statements* (Washington, 1986).

[7] U.S. Senate, Foreign Relations Committee, *Hearings on the INF Treaty*, 100th Cong., 2d sess. (1988), part 1, 289 (emphasis added).

ices.[8] The difference between the Reagan administration's definition of "effective" and the Carter administration's definition of "adequate" now seems imperceptible. The 1977 "adequate" standard is, at any rate, the only statutory reference.

We assume that "adequate" requires, as a minimum, that

1. no violation that could endanger national security should remain undetected and unidentified;
2. a violation should be identified in sufficient time to allow remedial action to protect national security; and
3. no violation that interferes in a basic way with the essential purposes of the treaty should remain undetected and unidentified.

As a practical matter, verification provisions, however strict, are unlikely to provide persuasive proof of violations. Even if a "smoking gun" indicating suspicious conduct is discovered, demonstrating an actual violation may be difficult. Rather, the role of verification provisions should be to deter violations by making the costs and risks of treaty evasion greater than the perceived benefits. No verification provisions can ever identify all violations with certainty. Conversely, no violations can ever be conducted at zero risk of being unmasked, since the "national" and "soft" methods of verification are presumably in place even without explicit provisions for treaty verification.

Thus, any agreed-upon verification regime represents a balance of values, including those reflected in the "adequate" standard as well as in the problems of intrusiveness and cost. The INF Treaty illustrates achievements of such a balance. The willingness of the Soviet Union to provide extremely detailed data and to accept on-site inspection as part of the INF Treaty has rightly been heralded as a dramatic and constructive departure from past positions. The INF Treaty, in Article XI, gives the "right to conduct inspections provided for by this article." These inspections are diverse and extensive.[9] However, they are specifically restricted to *declared* facilities. They include inspections to verify elimination and destruction

---

[8] U.S. Senate, Armed Services Committee, *Hearings on NATO Defense and the INF Treaty*, 100th Cong., 2d sess. (1988), part 1, 40–41, 50–51.

[9] For example, about 180 inspections are expected to be carried out within the first month after the treaty comes into force. Inspection rights in the allied "basing" countries are included, and the treaty also provides for cooperative measures such as removal of roofs over specified activities to enable overhead satellites to make observations.

of missiles covered by the treaty, to guard the portals of facilities in which the final assembly of ground-launched ballistic missiles is accomplished or has been accomplished, and to examine various supporting activities identified in the treaty. The protocol annexed to the treaty specifies the conduct of such inspections in great detail.

The omission of rights to inspection of any *undeclared* facilities is significant. Although the United States initially proposed inspections of undeclared facilities, the proposal was withdrawn once it proved nonnegotiable at home; there was concern on the U.S. side that such inspections might compromise sensitive U.S. technology. The Reagan administration, in its campaign to secure ratification of the INF Treaty, is correctly emphasizing the unprecedented extent of the inspections provided for in the treaty, but it cannot claim, in reply to the critics, that inspectors can go anywhere. Yet we believe the provisions fully meet the "adequate" standard cited.

A further factor that sometimes enters into the balance of values is precision of observation. Once quantitative determinations are required to verify compliance, one must understand that any measurement necessarily is subject to some lack of precision. This is true whether measurement is a count of permitted systems or an estimate of physical quantities. It is recognized, for instance, that counting of mobile missiles, which can be hidden in buildings or woods, is more uncertain than counting of fixed silos. One must balance the advantage to security of greater survivability of mobile missiles against the disadvantage of uncertainty of numerical verification.

Precision of measurement required for verification has been a recent contributor to two controversies relating to compliance.[10] One relates to the charge made by the Reagan administration that the Soviets violated SALT II in respect to the "one new type" limit and the other to verifying the nuclear yield limit specified in the Threshold Test Ban Treaty. In both cases, the disputes have been beset by confusion stemming from a lack of understanding of the uncertainty of physical measurement.

SALT II establishes rules specifying permitted variations in size and weight between a new missile and an old one, in order to determine whether a missile is or is not a "new type." The Reagan administration

---

[10]See *Compliance and the Future of Arms Control*, report of a project sponsored by the Center for International Security and Arms Control, Stanford University, and *Global Outlook*, Gloria Duffy, project director (Cambridge: Ballinger, 1988), 54–57, 62–74.

has charged that the SS-25 is a new type, whereas the Soviets claim that it is a permitted modernization of an old type, the SS-13. The issue involves problems both of measurement and of drafting. Means to measure dimensions and weight to the required precision of 5 percent are simply not available by NTM; and the information on the older SS-13 is particularly imprecise. Moreover, there is ambiguity in the treaty concerning whether or how the packages containing guidance, telemetry, and penetration aids are to be included in calculation of payload weight. The administration cannot substantiate its charge without greatly improved data nor validate it without a resolution of the dispute over the payload weight. In any case, it is difficult to maintain that the issue is one of major, or even significant military importance.

The Threshold Test Ban Treaty (TTBT) sets a ceiling of 150 kilotons for the permissible yield of an underground nuclear test explosion, and the issue of measuring that yield has received undeserved prominence. Verification considerations were not the determining factor in setting the yield threshold of the TTBT; from the point of view of detection and identification, the threshold could have been much lower. Instead, the threshold proved to be the minimum negotiable. In particular, the United States, considering its perceived national security needs, insisted that military needs required higher yield tests. Yet measurement of test yield near the threshold has in itself now become an issue.

A Reagan administration spokesperson states that national security "requires adoption of verification mechanism which will achieve a yield estimation accuracy of at least 30 percent with 95 percent confidence in order to find that the Threshold Test Ban Treaty is effectively verifiable."[11] The administration gives no evidence for how such a requirement relates to national security, nor does it make a credible case that the Soviet Union could exploit greater imprecision of measurement. Errors of measurement are of two kinds: systematic and random. Systematic errors occur if geological conditions at a test site systematically generate either high or low readings of earth motion. Random errors stem from the unavoidable dispersal of measurements due to such factors as uncertainties in the propagation of seismic signals and natural background noise. An evading party could exploit a detecting party's imprecision of measurement only through the knowledge of a bias in that measurement of which the detect-

---

[11]U.S. Senate, Foreign Relations Committee, *Hearings on the Threshold Test Ban Treaty and Peaceful Nuclear Explosions Treaty*, 100th Cong., 1st sess. (1987), 205.

ing party itself was unaware. If the uncertainty is purely random—that is, a result of uncontrollable variables—an observation may result in measurements that are either too high or too low. In view of this complex situation, it is absurd for the administration to assert that an uncertainty factor of 1.3 in explosive yield (an uncertainty in estimating yield of ± 30 percent) provides for "effective" verification, whereas a factor of perhaps 1.5 or 2.0 does not. An uncertainty factor of 2 would not mean that a violator could explode several devices of 300 kilotons with impunity. It would mean only that such explosions would have as good a chance of being registered at 150 kilotons as at 600 kilotons. The spectacle of experts arguing whether uncertainty factors of 1.3, 1.5, or even 2.0 are "adequate" or "effective" bespeaks a lack of understanding of the nature of measurement. Evidence is, in fact, persuasive that the Soviets have observed a yield threshold and that this threshold is consistent with a 150-kiloton limit. Thus, considering the uncertainties of physical measurement, no credible evidence of violation exists.

Implementing an Agreement. The third phase of arms control is implementation. No party will enter an arms-control agreement unless it believes that its national security will be greater with the treaty regime than without it. It is absurd to presume that the Soviets enter agreements under a "policy of cheating" to gain unilateral advantage, and there is nothing in the record to support such an assertion. Limited violations, ambiguous conduct, disputes rising from ambiguous provisions, and brinkmanship must be expected. The parties to an agreement must establish processes to prevent such action from producing significant military advantage for any one party and from eroding the treaty.

Both the United States and the Soviet Union have practiced brinkmanship. For example, the two countries have taken considerable risks by permitting radioactivity from underground nuclear tests to cross national boundaries, in violation of the 1963 Limited Test Ban Treaty. Activities on both sides "stretch the fabric" of the ABM Treaty.

Actual violations can, of course, differ widely in nature and origin. At an extreme, the highest level of a government may specifically sanction a real violation in order to seek unilateral advantage or to test the response and will of another party. Real violations, which may or may not be militarily significant, may also result from decisions made within a military bureaucracy without initial knowledge or approval from high-level authority, or at least without high-level awareness that violations may re-

particular—find it difficult to acknowledge error, let alone guilt. For reasons of national pride or just from inertia, all countries, under all systems of government, share an understandable reluctance to admit openly that they have violated international commitments. Thus, even if one party clearly establishes a violation by another (a rare occurrence), it must consider a wide range of responses. Deciding which of these options is appropriate depends, of course, on whether preserving the treaty regime remains in the responding party's national security interest. Assuming that the party wants to save the treaty regime, an appropriate response must signal that the violation is unacceptable but that compensation is possible within the treaty regime.

When negotiations to resolve a dispute over compliance do not produce a satisfactory solution, the offended party can publicize the violation in the hope that the offending party will halt its conduct (a method that seems to have been successful thus far in the case of the Krasnoyarsk radar). Or the offended party can acquire weapons to counter whatever military advantage the offending party might gain by its violation. Such a countermove may or may not itself violate the treaty. Finally, the offended party can terminate the treaty outright. Violating or terminating a treaty are responses of last resort, involving large political costs; but these responses are available to prevent real injury from noncompliance.

The governments of the United States and Soviet Union, acting together through consultative processes, should first seek agreement on how to interpret treaties in light of new problems and technologies and negotiate solutions to problems of questionable compliance. Assuming that the United States is not satisfied by these negotiations, it has further options. Within the U.S. government, there is often a debate, at this point, about whether a violation has been proved and, if so, what its consequence will be for U.S. national security. Those opposed to a treaty may be most vigorous in charging violations. At the same time, those interested in preserving the treaty usually realize that an agreement will gradually erode if violations do not draw a response but may collapse if the United States responds by violating the treaty in turn. The dilemma is not always easy to resolve. But, if diplomatic responses (through the SCC or otherwise) are unavailing, the United States can respond by compensating actions that do not violate the treaty, by violations proportional to those committed by the Soviet Union, or by withdrawal from the treaty. Violation or withdrawal would imply the deliberate conclusion that, in view of the Soviet Union's noncompliance, the treaty, as originally conceived, no longer served the national interest of the United States.

A Comparison of the International and Domestic Processes

The process of creating international treaties and ensuring compliance with their provisions has similarities and dissimilarities to the process of creating and enforcing domestic statutes. Because familiarity with domestic law tends to affect how we think about treaties, we sometimes loosely transfer the terminology, calling for "prosecution" and "punishment" of the "law breaker" when we believe an arms-control treaty has been broken. The Reagan administration's noncompliance reports read like the opening statement by a prosecuting attorney in a criminal case. In effect, the prosecutor represents the United States, the defendant is the Soviet Union, and the jury is the Congress and the public. Such a scenario is not a realistic view of the international compliance process.

The Drafting of Treaties and Statues. Both international treaties and domestic statues need two kinds of process to make them work: a law-giving process with the "consent of the governed" and a fact-finding and enforcement process, if such consent is insufficient to prevent disputes or infractions.

The lawgiver for U.S. domestic statutes is the legislature; our consent to be governed by these statutes is derived, in democratic theory, from the election of the legislature that represents us. There is no equivalent world legislature; the United Nations could not become a legislature for arms-control statutes for the United States and Soviet Union without the consent of the two superpowers—and they are not about to give it. Yet, it is by consent to the negotiated terms of an arms-control treaty that parties take the first step toward compliance. No party will agree to terms that it perceives are against its national interests, and the negotiation process gives parties control over those terms. Consent of the governed is far less theoretical in the case of consent of nations to an arms-control agreement than in the case of consent of individual citizens to a statute adopted by their representatives elected by majority vote.

Verification and Inspection. The verification process for international agreements has essential differences from the investigative process for domestic law, but much can be learned from a comparison. We have listed fact-discovery methods for international verification. Investigation for domestic law enforcement uses some of these same techniques—for example, electronic and photographic surveillance, informants, and detectives.

In the United States, as in other countries, there are barriers to inspection anytime or anywhere for domestic law enforcement. The Fourth Amendment to the U.S. Constitution guarantees the "right of the people to be secure in their persons, houses, papers and effects, against unreasonable searches and seizures"; no warrants for *unreasonable* searches in such places may be issued "but upon probable cause." The vast body of domestic law interpreting this provision permits few searches by law-enforcement inspectors, unless they first obtain judicial warrants or administrative substitutes, for which they must show probable cause, except in case of an emergency, such as a fire, or of consent, real or implied, by the inspected party. Homes, of course, receive the greatest constitutional protection. Commercial buildings receive less; buildings of a business heavily regulated for health and safety reasons receive even less; open lands belonging to such a business receive even less still.[12]

In the arms-control context, on-site inspections anytime or anywhere by Soviet inspectors in the United States may confront constitutional objections about unreasonable searches if they include property not under government contract or control. The INF Treaty avoids the objections because only facilities declared in the treaty may be inspected, and they are all government-owned or subject to government contract. The government can ask defense contractors to give prior consent to inspections as a condition of further contracts.

For START, however, Gorbachev and Reagan agreed at the 1987 Washington summit that undeclared locations could be inspected. For constitutional and security reasons, however, the inspections are not to be anytime or anywhere. The joint U.S.–Soviet summit communiqué calls for a treaty giving both sides "the right to implement, in accordance with agreed-upon procedures, short-notice inspections at locations where

---

[12]A recent U.S. Supreme Court case permitting warrantless inspection dealt with an open commercial quarry subject to health and safety regulation pursuant to federal statute. The statute, which was upheld, permitted the owner, however, to resist forcible entry by the inspector, who then had to go to court to make a showing of probable cause before he could enter (Donovan v. Dewey, *United States Reports* 452 (1981): 594).

either side considers covert deployment, production, storage or repair of strategic offensive arms could be occurring."

Within the short-notice period, either side can make a showing of probable cause that the treaty requirements for inspection have been satisfied. Officials from the inspected country will certainly meet inspectors at the border to deal with local authorities and to prevent conduct not permitted by the treaty. Thus, officials from the inspected country can take the role of judges for a showing of probable cause. For inspections of defense industries or weapons deployment sites, the constitutional problem is either nonexistent because of government ownership or can easily be dealt with by requiring contractor consent. But, in the case of a farmer's cattle barn large enough to hide a mobile ICBM, some showing of probable cause would be necessary. However, it is impossible to specify precisely the extent of warrantless inspection that the courts will consider is constitutionally permissible under an arms-control treaty.[13]

Security problems also prevent agreement through START to inspection anytime or anywhere. For example, the United States does not want Soviet inspectors to find out about stealth technology, or to inspect Pentagon war rooms. The Soviet Union obviously faces similar security issues. The 1987 summit communiqué not only requires notice but also "agreed-upon procedures" to deal with such problems. What sort of probable cause and security safeguard procedures must be written into a treaty calling for on-site inspection of undeclared facilities?

For some kinds of suspicious activities, the treaty definition of probable cause for an inspection by one side may automatically exclude what the other wishes to keep secret. For suspected underground tests of nuclear weapons, for example, the definition of probable cause would describe the size and characteristics of earth tremors detected at long distances and would specify for inspection a circumscribed area around the center of suspicious tremors. Underground tests likely would not be conducted near such secret facilities as stealth laboratories. Thus, in practice, an area circumscribed for inspection would probably include few sensitive facilities.

In the case of inspection for mobile missiles that can be hidden in woods or a large shed, the problem is more difficult. The treaty, in addi-

---

[13]See Thomas A. Connolly, "Warrantless On-Site Inspections for Arms Control Verification: Are They Constitutional?" *Stanford Journal of International Law* 24 (1987): 179; and Edward A. Tauzman, "Constitutionality of Warrantless On-Site Arms Control Inspections in the United States," *Yale Journal of International Law* 13 (1988): 21.

tion to defining the kinds of objects inspectors may look for, can also define exclusion areas that they may not inspect. Alternatively, in the interest of protecting sensitive facilities, the treaty can define areas within
which mobile missiles may be deployed and inspected. But such restrictions must be established on a reciprocal basis. Thus, in the interest of
protecting sensitive facilities, the treaty may unavoidably create a gap in
the inspection coverage of undeclared facilities.

Quite apart from the complexity of the legal and security factors, the
degree of acceptable inspection becomes a matter of political judgment.
Excessive inspection rights will provoke opposition from those resenting
the presence of foreign intruders as well as from guardians of civil liber-
ties; insufficient inspection rights will give the opportunity for treaty opponents to cite loopholes for evasion. One can expect that, in practice,
both the United States and Soviet Union, in an effort to balance these factors, will be moderate in demands for inspection, each recognizing that
the other will reciprocate.

Maximizing Compliance. Maximizing compliance with an arms-control
treaty is essentially different from enforcing a domestic statute. In the
case of a domestic statute, if there is serious disagreement about relevant
facts or law, one can have recourse to a third party (a jury, judge or
administrative agency) to arbitrate, to order enforcement, or to punish the
violator. Nothing similar is available for arms-control treaties. No
sovereign country has yet been willing to delegate decisions to a third
party on issues as important to its security as compliance with a treaty
governing weapons or weapons tests. The closest the international system
comes to third-party verification is the International Atomic Energy
Agency (IAEA), which monitors civil nuclear facilities at declared
locations. Most of these facilities are located in countries that have agreed
not to acquire nuclear weapons. The purpose of IAEA inspections is to
see that products from a civilian nuclear-fuel cycle are not diverted to
military use. But few countries are likely to give the IAEA permission to
search for undeclared facilities, to impose punishment, or to render final
judgment. Its present practice is only to inspect and report back, not to
make roving inspections or to judge guilt or innocence.

A better domestic analogy for arms control than criminal law would be
a hypothetical agreement, among Chrysler, Ford, and General Motors not
to compete by cutting costs for auto safety. Let us imagine that Congress
has not exercised jurisdiction over auto safety (as in fact it has) and that
the Big Three want to keep the government out of that field. Let us as-

sume that the automakers negotiate their own auto safety standards but agree never to resort to a court to enforce them, because the government might perceive the agreement not to compete by cutting costs for safety as a violation of antitrust laws.

What could Chrysler do if it decided that Ford was cheating on the agreement in a new model, utilizing new technology not covered by the standards? It could abrogate the Agreement. But if the standards still seemed of value, that would be against Chrysler's interest. It could cheat on the standards itself in a similar way. But that would likely stimulate the others to follow suit, and eventually the agreement would be destroyed. It could call the three together in an attempt to get Ford to stop or, alternatively, to get agreement of the three on how to handle the new technology. Indeed, if the three foresaw that changes in technology and vigorous competition were likely to produce disputes of this kind, they could provide for periodic private meetings of expert representatives from each company as part of the basic agreement. The representatives would review compliance problems and try to work out solutions.

The analogy to arms control is clear. The United States and the Soviet Union agreed in SALT I, SALT II, and now the INF Treaty to utilize consultative commissions. These commissions are composed of representatives of the two countries, who meet privately, acting on instructions from their governments, to deal with compliance problems.

Neither the international consultative bodies nor the hypothetical domestic arrangements among Chrysler, Ford, and General Motors can apply sanctions. Thus, the recourse available to the United States to achieve compliance with an arms-control treaty is similar to that available to Chrysler in the hypothetical case. Assuming the treaty is still in the interest of the United States, despite a question of violation, the goal should be to deter future violations while preserving the treaty. The goal—deterrence of violations—implies maximizing incentives for compliance.

Because of the absence of an independent adjudicating party charged to enforce compliance with arms-control treaties among sovereign nations, the process cannot be the same as for enforcement of domestic laws.[14] However, sanctions by the parties to a treaty can be effective, with

---

[14]For discussions of promoting compliance under an arms-control treaty, see Fred Charles Iklé, "After Detection—What?" *Foreign Affairs* 39, no. 2 (1961): 208; Carnegie Panel on U.S. Security and the Future of Arms Control, *Challenges for U.S. National Security*, Final Report (Washington: Carnegie Endowment for International Peace, 1983), 53–54; and Compliance and the Future of Arms Control, 12–15, 198-209.

the prerequisite that they find the treaty still to be in their interest. A domestic law continues to govern the behavior even of those who no longer accept it. However, parties to an arms-control treaty cannot compel each other to comply with the treaty if one or the other has decided that compliance is no longer in its interest. If a party wishes to shed the constraints on arms competition imposed by the treaty and is willing to accept whatever sanctions and political consequences may arise, it cannot be stopped from withdrawing from the treaty, from violations, or from reinterpreting the treaty to its advantage.

Assessment of Compliance

Given the differences between the processes for ensuring compliance with arms-control treaties and domestic statutes, what are the differences between the expected degrees of compliance with new treaties and new statutes? In enacting statutes, legislators are unlikely to worry too much about the level of compliance. Rather, they are likely to consider such questions as what effect the statute will have on the public or industry, and which of their constituents favor or oppose it. When they think about compliance, they are likely to do so on the basis of a net assessment, asking, for example, what will be a statute's expected net impact on society, assuming a typical level of compliance for such a statute.

Looking at the entire record of compliance with arms-control agreements rather than just at the disputes that have arisen, we can state unequivocally that the superpowers have done well. Indeed, the record compares very favorably with that for compliance with domestic legislation in this country. Many murders occur each year despite murder statutes, and the drug problem is rampant despite antidrug legislation. Although legislators worry about such noncompliance, their likely response is to provide more money for enforcement or for social or economic measures designed to decrease motivation for crime. They would hardly repeal the murder and drug statutes and permit the crimes to go legally unchecked. Massive noncompliance may produce statutory changes, as in the case of the repeal of prohibition and the partial repeal of the 55-mile-per-hour speed limit on national freeways. But, in such cases, the reason for change is not so much the record of violations as the basic unpopularity of the legal restraints.

Net assessment of a proposed arms-control regime, in a manner similar to the net assessment of domestic statutes, should compare the expected robustness of national security with or without the agreement. We invoked the standard that an arms control agreement is worthy of support by the United States if, with adequate verification, the resultant regime improves U.S. national security. From the U.S. point of view, such an assessment must analyze the regime that might result over a range of possibilities from one extreme, full compliance by both sides, to another, full compliance by the United States but substantial noncompliance or abrogation of the treaty by the Soviet Union. The assessment must weigh the military and political impacts over this range of possibilities and the responses each side has available to the other's action. The goal is to determine whether or not the arms-control agreement will achieve greater national security at lesser risk and burden of armament.

A Tale of Three Treaties. What does a net assessment of three treaties—the ABM Treaty, the SALT II Treaty, and the Threshold Test Ban Treaty—show in the light of the actual record of compliance?

Each of the treaties has a different legal status. The first agreement is in force for an unlimited period but subject, with six months' advance notice, to the right of withdrawal by either side for reasons of "supreme national interest." The second was never ratified by the United States, was repudiated by the Reagan administration, and now has lapsed. The third also has never been ratified by the United States, but, thirteen years after signature, U.S. negotiators are pressing for additional verification measures as a condition of ratification. The treaty has not been repudiated.

We have touched on the compliance record as well as some of the disputes of interpretation and conduct associated with these three treaties. Yet all three agreements *have, in fact, effectively governed* the conduct of the signatories and still do. For instance, with minor exceptions, the United States and Soviet Union are still following the principal numerical limits of SALT II on strategic delivery vehicles, in part as a result of action by Congress after the administration repudiated SALT II. Both side have deactivated systems to meet treaty terms. As to the ABM Treaty, neither the Soviet Union nor the United States has so far provided a base for territorial defense against ballistic missiles; and thus both countries— as well as the United Kingdom, France, and China—are assured of the effectiveness of their deterrent forces for at least a decade, if not decades, ahead. In the case of the Threshold Test Ban Treaty, there is no credible evidence that either of the superpowers has conducted any nuclear test with a yield of more than 150 kilotons.

These three treaties have provided predictability for budgeters and military planners. Their provisions have limited conduct and tempered the worst-case projections that intelligence analysts on both sides inevitably make. The treaties have accomplished their intended purposes during and even beyond their anticipated lifetimes. We believe the treaties have improved U.S. national security.

During all of the post–Korean War history, there has been only one instance in which the evidence of a violation has been persuasive: the Soviet Krasnoyarsk radar. Construction of the Krasnoyarsk installation started in the late 1970s, continued at a slow pace, and has now stopped. In its technology and even many details of its dimensions, it is an exact copy of large phased-array radars constructed for early warning of ballistic-missile attack—in particular, the Pechora installation to the northwest. These early-warning radars are located, as the ABM Treaty requires, on the perimeter of the Soviet Union looking outward. The Krasnoyarsk radar is, however, located inland by hundreds of miles, at the end of available railroad lines, and on the edge of the permafrost, where technical construction is difficult and expensive. The radar overlooks several thousand miles of Siberian landmass. It thus appears clearly to be an early-warning radar that violates the ABM Treaty because it is not "along the periphery" of Soviet territory and "oriented outward" (Article VI (b)).

The Soviets initially countered accusations concerning the Krasnoyarsk radar by claiming disingenuously that it would serve a legal space-surveillance mission. However, considering the space it could cover, the radar would be ill suited for that mission alone. The Reagan administration claimed that the Soviets were signaling an intent to break out of the ABM Treaty by starting to provide a base, forbidden by Article I of the ABM Treaty, for a defense of their territory. The administration made assertions, unsupported by technical evidence, that the Krasnoyarsk radar would have a battle-management capability and, by the early 1990s, could attain significant territorial ABM capability. However, the evidence indicates that there is no adequate infrastructure linking this radar to other installations; and work on other legal ABM projects—in particular, the modernization of the Moscow system—has been excruciatingly slow. Also, the technical characteristics of the Krasnoyarsk radar, especially its operating frequency, are such that, in the presence of nuclear explosions, even if the radar were not totally blacked out, it would observe grossly distorted missile trajectories.

It is interesting to note that the United States had knowledge of these technical factors even before the Soviets agreed to have a congressional delegation inspect the Krasnoyarsk radar. In fact, the congressional "on-site inspection," although a welcome political gesture, did little to illuminate the issues.

U.S. Approaches to Soviet Noncompliance. The Reagan administration has been selective in deciding when to be confrontational about Soviet noncompliance. In the case of treaties it doesn't like, it has been confrontational, as the preceding discussion shows. But it has adopted a problem-solving approach in dealing with agreements it does like. In a speech to the United Nations in the fall of 1985, President Reagan chose to praise only the nuclear Non-Proliferation Treaty (NPT) among the several arms-control treaties then in effect. Officials of his administration met regularly with counterparts from the Soviet government to talk about compliance with that treaty. The United States was concerned about a nuclear reactor in North Korea for which the Soviet Union may have provided assistance or nuclear materials. The reactor was large enough to make a significant amount of plutonium, which, with reprocessing, could be used to make nuclear weapons. The reactor was not subject to international inspection because North Korea was not a party to the NPT; nor had that country accepted inspection of the reactor in any other way. It would be a violation of the NPT for the Soviet Union to provide nuclear materials or assistance to a nonnuclear country not a party to the NPT for a reactor not to be subject to inspection.

But the administration made *no charge* in the public noncompliance reports. Instead, administration officials complained privately in their regular talks with their Soviet counterparts. Some time later, apparently as a result of the talks, North Korea joined the NPT; as a result, its reactor is now subject to international inspections, and the U.S. complaint has been resolved.

The administration also values the U.S.–Soviet "Incidents at Sea" Agreement, which enjoins naval ships and aircraft of the two countries from "playing chicken" with each other. The purpose of this significant arms-control agreement is not just to prevent accidents but also to minimize risk of a U.S.–Soviet war triggered by overcompetitive naval forces. After the Soviets shot down the South Korean airliner on flight 007 in

1983, Soviet naval ships used shouldering and other tactics against U.S. and other naval vessels looking for the sunken aircraft's black box or other evidence or remains. On the basis of newspaper stories, such tactics appeared to result in many violations of the "Incidents at Sea" Agreement. But the administration made *no charges* to this effect in the public noncompliance reports. Instead, it made private complaint to the Soviets, and the noncompliant conduct generally stopped. Such consultation with the Soviet Union has been at least as effective as self-righteous public accusations in achieving compliance.

In respect to negotiations on restraining nuclear tests, the disputes have been ostensibly over verification. In actual fact, differences on the importance of continued nuclear testing to national security are now blocking agreements on further limits on nuclear tests. Negotiations on cessation of nuclear tests between the United States, the Soviet Union, and the United Kingdom opened after the 1958 Conference of Experts had examined technical verification issues. Agreement in the 1960s on nuclear test bans appeared to founder on differences between the Soviet Union and the United States on the nature and number of on-site inspections that might be permissible; reportedly, the two parties differed in the end on whether the inspections should number three or seven, respectively. Lack of adequate precision to measure the threshold is the nominal issue holding up ratification of the Threshold Test Ban Treaty, signed in 1974. The current Soviet–U.S. dialogue on nuclear test matters focuses on cooperative verification, rather than on the treaty restraints themselves.

The Reagan administration has been candid in announcing a policy that nuclear testing should continue as long as nuclear weapons are the key element in deterring war.[15] Whether one does or does not agree with the judgments expressed in that policy, we note that there is no pretense that verification is any longer the issue.

The Overall Compliance Record. We have enumerated a number of examples that have been cited in the series of presidential reports entitled *Soviet Noncompliance with Arms Control Agreements,* the latest released

---

[15]"Nuclear weapons will remain the key element of deterrence for the foreseeable future. During such a period, where both the United States and our friends and allies must rely upon nuclear weapons to deter aggression, nuclear testing will continue to be required. A carefully structured nuclear testing program is necessary to ensure that our weapons are safe, effective, reliable, and survivable" [U.S. Department of State, *U.S. Policy Regarding Limitations on Nuclear Testing*, Special Report no. 150 (Washington, August, 1986)].

December 2, 1987. [Further reports in this series, the latest in 1992—even after there was no longer a Soviet Union—were released by the Reagan and Bush administrations after the completion of this Working Paper.] We discussed the most important challenge to Soviet compliance, the Krasnoyarsk radar, in some detail. We conclude that the clear purpose of that radar is to fill the gap in ballistic-missile early-warning coverage created by the operating ranges of the U.S. ballistic-missile submarines. Both the physical situation and informal Soviet contacts support our conclusion. Krasnoyarsk does not constitute a military risk. The radar is an installation that is dedicated to a function that is legal under the ABM Treaty; but its location is illegal. The most likely interpretation of the Krasnoyarsk violation is that the Soviets made a decision, possibly not at the highest level, to give priority to economic concerns over treaty obligations.

The question still remains how the Krasnoyarsk issue and related issues concerning the U.S. radar at Thule, Greenland, and the one planned for Fylingdales Moor, England, should be handled. As newly constructed radars, these installations would clearly be illegal under the ABM Treaty; but the U.S. administration justifies them as "modernizations" of obsolete radars built in the vicinity before the ABM Treaty entered into force. Ultimately, such issues might have to be resolved by agreements analogous to "consent decrees," by which parties agree not to repeat certain practices but do not explicitly admit guilt in connection with past practices. All other cases of noncompliance alleged in the presidential documents either have been resolved or are, in fact, disputes over ambiguous conduct or treaty provisions.

Thus, overall, *the record of compliance has been good.* We believe that, despite a few problems, a net assessment shows that the arms-control treaties we have discussed have improved U.S. national security. Disputes relating to conduct or interpretations of provisions are *inherent* in the arms-control process, as they are in the enforcement of domestic law. Negotiators recognized this fact when they provided, in the ABM Treaty, for the Standing Consultative Commission and, in the INF Treaty, for the Special Verification Committee and the creation of the new Nuclear Risk Reduction Centers. Such consultative bodies can continue to be viable only if the parties utilize them in a problem-solving (rather than confrontational) spirit and if they refrain from public accusations until the processes of these consultative bodies and other diplomatic channels have been exhausted. We do not suggest that violations of arms-control agreements should be condoned but, rather, that questions about compliance should be expected. But such questions need not be synonymous with violations.

The current discussions on the merits of ratifying the INF Treaty and moving on toward a START agreement really concern the strategic military merit of the elimination of the classes of delivery vehicles considered. It is no accident that those who question the effectiveness of the *verification* provisions are often those who fundamentally oppose *elimination* of these delivery vehicles.

We hope that our discussion induces the reader to view compliance issues in a new perspective. The heart of the matter is how arms-control restraints affect national and international security.

Science Advice at the Presidential Level

Much has been said and written on the need for expertise in science at the White House level. Recently there appeared a book entitled *Science and Technology Advice to the President, Congress and Judiciary,* edited by William Golden, containing a large number of essays by people who had participated in the advisory process. During the 1988 presidential campaign, several scientific societies submitted questionnaires on presidential science advising to the candidates, but neither has made this a public issue. Yet many fateful decisions resulting from the interaction of man and nature—from the ozone layer, acid rain, and the greenhouse effect to the weapons of war—will have to be made by the next president. He can disregard scientific input only at greatest peril to the nation and future generations.

Most governmental departments have scientific staff and science advisory bodies and also have scientifically competent "think tanks" under contract. Members of Congress have scientifically competent staff members, and the Office of Technology Assessment, under a bipartisan directorate, responds to requests generated by congressional committees. Yet the structure of advice to the president remains unresolved. I maintain that there is a paramount need for the president to have access to impartial, independent, and informed scientific advice that has not been filtered through the policy objectives of governmental departments or agencies.

There are diverse functions that can only be satisfied at a presidential

A talk given at Miller Center, University of Virginia, September 24, 1988.

level. In broad terms, scientific advice to the president can be divided into two categories: (1) "science in government"—that is, scientific input into policy making—and (2) "government in science"—that is, the management of that part of the scientific endeavor for which the government has responsibility. I consider it to be most important that the former be carried out at the presidential level. For a rational decision-making process, the president should have the best of scientific counsel available before he makes policy decisions on issues that have a technical component.

The management of science by government and the question of priorities and budgets for science support is less in need of consultation at the presidential level. The present pluralistic system of support of science by government through diverse agencies and the regular budgetary process headed by the Office of Management and Budget has served American science well, notwithstanding the obvious shortcomings of the process. Scientific priorities among competitors for the government dollar are impractical to resolve by a single council of scientists, however wise and broad-minded. Scientific advice dealing with broad policy issues affecting the level of support of the different segments of science—basic versus applied, big versus small, specific "super" projects—is needed, however, at the highest level. The normal process is that the departments and agencies formulate their programs through peer review of the proposals received. In turn, the departments and agencies interact with the Office of Management and Budget. The office of the Special Assistant to the President for Science and Technology can and should interact with OMB, but only on a selective basis when critical policy matters are at issue. The National Academy of Sciences can play, and has played, a very valuable role in ordering science priorities by carrying out periodic reviews on the status and opportunities of the various sciences. These studies in general recommend a "maximum reasonable" level of governmental science support—that is, a level consistent with the ability of the scientific community to carry out the programs responsibly. Such studies, together with realistic budgetary constraints, give useful guidelines for actual program decisions.

There is another reason why presidential science advice should concentrate on the science component of governmental policy rather than on support of science by the government. Opponents to a science presence at the White House level frequently claim that this creates a built-in "science lobby" right within the White House, and that this, in turn would distort budgetary priorities. If science advice at the White House level largely stays clear of details of governmental support of science, this criticism can be mitigated.

Any advisory system serving the president must be *selective*. If a White House system were to review all science-related decisions, it could only become another level in a giant bureaucracy, and no one wants that. But if the mechanism is to be selective then the question is, who selects? Any science-advisory system in the Executive Office of the President must serve the president rather than represent the scientific community. That means it must be responsive to requests by the president or his staff for analysis and counsel in time adequate to the need. However, a science-advisory system at the presidential level must also serve as an "early warning" system to alert the president when action is needed, even if the president has not been conscious of such a need. Thus, the agenda for a selective presidential advisory system must have room both for items generated by the advisors and by the advisee—in this case, the president.

There is only one president, and his agenda reflects his priorities among a wide variety of needs. Thus, one cannot expect that a presidential science-advisory system requires frequent personal communication with the president. However, it is essential that such a personal presentation is, in principle, possible and that is does occasionally occur. Even if the advice is, in fact, received only by presidential staff and, through them, by government agencies, the fact that the advisory system *can* reach the ear of the president gives a great deal more weight to such counsel.

This, then, raises a question: How many voices of science shall reach the president's ear? This is a matter that cannot be legislated or determined without reference to the individual president's interest or scientific literacy. The number, depending on each president's inclination, can vary between zero, if he is simply not receptive to direct scientific input, or a substantial number, if he would like to be exposed to a variety of views. The most popular number with past presidents has been *one*. This implies that one individual must wear a variety of hats, in addition to his primary role of science advisor to the president. The task of coordinating the work of the different agencies that is relevant to science has been carried out under various administrations through a Federal Council of Science and Technology, with the chairman of that council generally being the one scientist who has access to the president. As director of the statutory Office of Science and Technology Policy (formerly the Office of Science and Technology), that individual is accountable to the Congress. As chairman of outside presidential advisory committees, he or she must convey the output of such bodies to the president.

If only a single voice dealing with "science" reaches the president's ear, there are dangers. Scientists like to pat themselves on the back for their scientific objectivity. Yet little science advice deals with recital of well-established scientific fact, although tutorials to update the president on current science have occasionally been given. Advice is most sorely needed when facts must be heavily mixed with judgment and when circumstances demand a great deal of prognostication of future developments in science and application of those developments to public affairs. Thus, most of science advice is overtly or inadvertently affected by the ideology and general outlook of the advisor. Even if only one scientific voice communicates with the president, that voice must be backed up by a structure through which a diversity of views can be aired, and it would be advisable if, in exceptional circumstances, members of that "back-up" organization had access to the president. Therefore, in addition to the science advisor, some of the inside and outside persons supporting the advisor should also be presidential appointees.

The back-up to the president's science advisor should consist of two tiers—a professional, full-time, in-house staff and a part-time group of professional, truly independent scientific advisors. Myraid inputs directed to the chief executive require scientific competence. There are detailed budgetary matters, there is simply the matter of "answering the mail," there is Congressional liaison, and broad policy must be applied to innumerable actions to be executed by the various governmental departments. All of this requires scientific technical staff of high competence in the executive office. Yet such a staff must not assume the role of *line* directors or managers of specific government action.

Then there is the matter of independent outside advice. The President's Science Advisory Committee (PSAC), established under President Eisenhower, is well known, as is its abolishment by President Nixon, and the establishment of the White House Science Council (WHSC) under President Reagan, which advises the science advisor but rarely the president. Why has there been so much controversy about the merits or demerits of these organizations? At issue is the search for competent and knowledgeable, yet independent, advice. Competence and detailed knowledge are most often acquired from direct responsibility for a program or function, and this may conflict with the desired level of indepen-

dence. But if advisors are truly independent, then the in-house professionals often challenge their competence and knowledge. Thus, the choice of
scientific counsel, by its very nature requires a balance of values.

The usefulness of any form of advice by a scientist to a lay person depends at some level on "credibility." This is an intangible quantity, and, in
the case of presidential science advice, it depends on the personal relationship between the science advisor and the president. No organizational
arrangement can change this fact. Substituting a multiplicity of voices for
a single advisor generally does not help, since a president cannot easily
adjudicate conflicting views if each side maintains that his advice is
based on science and is backed by credible analysis. When witnesses testify before Congressional committees and when the staff arranges for adversary testimony by scientists, the lay congressman is faced with the
problem of "which liar to believe." Thus, unavoidably, the effectiveness
of a presidential advisory system depends not only on the personal relationships but also on the prominence of the advisors; such prominence
usually has been attained in connection with topics not related to the sub-
jects of government advice. Prominent advisors can be attracted into the
presidential advisory circuit only if they have some confidence that their
role will be truly useful. Thus, although the actuality of access and the re-
porting chain within the White House may not be of crucial importance
as such, the *potential* of direct access to the president and a direct presidential appointment is desirable to attract the level of talent required for
broad credibility of the advice given.

There are many lessons to be learned from the operation of PSAC and
the science advisor. First there is the matter of confidentiality in the relationship between the president and his advisors. PSAC has been accused
of viewing its role more as a national advisor than as part of the president's family. It has also been accused of being unwilling to accept presidential decisions and "going public" to counteract such decisions.

Such charges are factually incorrect; they are based on a confusion between the role of PSAC and the role of its individual members. There is
indeed a fundamental problem. When prominent individuals constitute a
high-level committee whose purpose is to advise the president, then a
question may arise about the extent to which the president "owns" such
individuals, even though they serve him only part time. PSAC has had an
excellent record in maintaining confidentiality about the advice it has
given to the president and about its results. There were no leaks. However, individual members of PSAC, during their tenure as presidential ad-

visors, were also asked to testify before Congress, to give advice to other executive agencies, and to speak in public forums on national issues. Clearly, it behooves any individual member of PSAC in such a situation not to divulge the nature of the advice given by PSAC to the president or in any way to compromise information that he received in his role as a member of PSAC. However, this has not been a problem in practice. The difficulty arises either when advice given to the president by PSAC has not been unanimous or when the president has rejected such advice. Clearly, the president can make a decision contradicting the advice, taking into account other, generally political, factors. In such a case, a member of PSAC, speaking as an individual, may express an opinion, in a different forum, that is in conflict with the president's decision. Yet it is known that he is a member of PSAC. Is this a permissible situation?

The key event in President Nixon's decision to abolish PSAC was that, one year after he decided to go forward with government support of a commercial Supersonic Transport (SST), a member of PSAC testified against the SST in Congress. Although his testimony did not explicitly state that PSAC had advised the president against the SST, the fact that he had been the head of a small ad hoc panel on the SST in the Office of Science and Technology created the public perception that PSAC had advised the president on that matter and that the advice was negative. The consequence was the perception by some that "PSAC could not be trusted"; yet what was under fire here was the expressed opinion of an individual member of PSAC clearly stating that he was speaking as an individual. In fact, the member had long been head of the military aircraft panel for PSAC, and, as such, had reviewed the SST program for President Johnson in 1965.

Judgment on the extent to which individuals serving as part-time presidential science advisors should exercise their right to give advice to other agencies is a matter of balance of values. If carried out to excess, then indeed the presidential advisory system will suffer. However, if such expression of personal views were enjoined entirely, very few prominent individuals would be willing to accept such an advisory role.

The president is expected to accept some advice while rejecting other advice. Therefore, it should be unreasonable for the president to expect the science advisor to become a public spokesman promoting presidential policies whether they are or are not in conflict with the scientific advice received. The role of advisors differs here relative to that of line officers of the government. Line officers, in particular those serving in the cabinet, indeed must support presidential decisions or, if they cannot in good conscience do so, resign. However, presidential advisors, even though they are indeed part of the White House "family," can serve the president best if they are not at the same time spokesmen for presidential policy. There is no problem if the president's science advisor publicly supports presidential decisions dealing with "government in science"—that is, support of science by the government—if he participates in purely ceremonial functions as a representative of the president, or negotiates for the president on matters of scientific cooperation with foreign nations, and so on. However, when the president has made policy decisions in foreign or domestic policy after receipt of science advice but where the primary policy topic is not science, the science advisory should best be silent. What was probably the fatal blow to the presidential science advisory system was struck when the president's science advisor, at the request of the Nixon administration, became a public advocate for the "Safeguard" antiballistic missile system then proposed.

Another important issue is the role of the presidential science advisory system in national security. Here independent scientific advice is most urgently needed. Indeed, the Defense Department supports a large number or scientists and utilizes a plethora of scientific advisory committees in its decision-making processes. A multitude of "think tanks" that carry out security studies commonly have large scientific or technical components. Yet such studies can rarely be viewed as fully objective. The majority of science advisory bodies in the Defense Department are composed of defense-contractor personnel. Few reports are released before comments from Defense Department officials have been accommodated by the report. Thus, independent science advice dealing with national security issues is sorely needed at the presidential level, unfiltered through the policy objectives of any one agency. In principle, the National Security Council (NSC) could have its own science staff, but this has not, in fact, been the case. The National Security Council's function is to coordinate inputs from the different parts of government concerned with national security, such as the Defense Department, the Arms Control and Disarmament Agency, the State Department, the Joint Chiefs of Staff, the intelligence agencies, and at times, the Department of Energy. Either the NSC induces consensus or it identifies interagency differences for presidential decision. The national security advisor, who chairs the NSC process, has seldom been sensitive to scientific and technical issues, to the detriment of the quality of the decision process.

In the old days of PSAC, the inverse problem existed. The agenda of PSAC tended to be pre-empted by national security items, including arms control, so that consideration of other science policy items was frequently slighted. The solution to this problem was to have the presidential science advisory system promote an increase in scientific competence in all government agencies, both within and without the national security arena at the policy level. For instance, the positions of Assistant Secretary for Research and Development and Undersecretary for Defense Research and Engineering were created at PSAC urging. It is unfortunate that, in recent times, this process has been reversed: for instance, in the Defense Department, the top technical officer no longer operates at the undersecretary level. In addition, during the PSAC days, joint appointments were made to the NSC and the president's science advisor's staff.

What do all these general considerations amount to? Am I advocating return to the old PSAC system, or am I satisfied with the less prestigious White House Science Council now in being, or am I suggesting some other method? My conclusion after weighing the general considerations covered here, is that PSAC chaired by the science advisor is, on balance, a better model. However, the preceding discussion highlights only some particular issues. Clearly, there cannot be a universal prescription for the "optimum" science advisory position at the presidential level. The individual preferences and tastes of the president must ultimately decide.

Index

About the Author

Wolfgang K.H. Panofsky has made far-reaching contributions to high energy physics. He determined the parity of pi mesons, showed that a neutral pi was lighter than a charged pi, and determined accurate pion masses. From this work, a collaboration with Lee Aamodt and Jim Hadley, the "Panofsky ratio" entered the literature. Through his initiative, the first information on the electromagnetic structure of the unstable excited states of the nucleon was discovered. He also contributed to the solution of many accelerator design problems.

Dr. Panofsky was born in Berlin, Germany, in 1919, and became an American citizen in 1942. He taught physics at the University of California at Berkeley and at Stanford University. He was a principal founder of the Stanford Linear Accelerator Center and served as its director for more than 30 years.

An early advocate of nuclear arms control, Panofsky played a prominent role in arms negotiations with the Soviet Union, as an advisor to President Dwight D. Eisenhower. In particular Panofsky served as chief delegate to the negotiation on detection of nuclear explosions in space. These talks formed the basis for the groundbreaking atmospheric test ban treaty signed during the Kennedy administration. In 1969 he was again at the forefront of the national debate over ballistic-missile defenses—an effort that contributed to the ABM treaty of 1972.

Dr. Panofsky has received numerous awards and honors, including the National Medal of Science, the Franklin Medal, the Ernest O. Lawrence Medal, the Leo Sztlard Award, and the Enrico Fermi Award.